Graduate Texts in Physics

For further volumes:
www.springer.com/series/8431

Graduate Texts in Physics

Graduate Texts in Physics publishes core learning/teaching material for graduate- and advanced-level undergraduate courses on topics of current and emerging fields within physics, both pure and applied. These textbooks serve students at the MS- or PhD-level and their instructors as comprehensive sources of principles, definitions, derivations, experiments and applications (as relevant) for their mastery and teaching, respectively. International in scope and relevance, the textbooks correspond to course syllabi sufficiently to serve as required reading. Their didactic style, comprehensiveness and coverage of fundamental material also make them suitable as introductions or references for scientists entering, or requiring timely knowledge of, a research field.

Series Editors

Professor Richard Needs

Cavendish Laboratory
JJ Thomson Avenue
Cambridge CB3 0HE, UK
rn11@cam.ac.uk

Professor William T. Rhodes

Department of Computer and Electrical Engineering and Computer Science
Imaging Science and Technology Center
Florida Atlantic University
777 Glades Road SE, Room 456
Boca Raton, FL 33431, USA
wrhodes@fau.edu

Professor Susan Scott

Department of Quantum Science
Australian National University
Science Road
Acton 0200, Australia
susan.scott@anu.edu.au

Professor H. Eugene Stanley

Center for Polymer Studies Department of Physics
Boston University
590 Commonwealth Avenue, Room 204B
Boston, MA 02215, USA
hes@bu.edu

Professor Martin Stutzmann

Walter Schottky Institut
TU München
85748 Garching, Germany
stutz@wsi.tu-muenchen.de

Marc Eichhorn

Laser Physics

From Principles to Practical Work in the Lab

 Springer

Marc Eichhorn
Institute Saint-Louis
Saint Louis, France

ISSN 1868-4513   ISSN 1868-4521 (electronic)
Graduate Texts in Physics
ISBN 978-3-319-37459-8   ISBN 978-3-319-05128-4 (eBook)
DOI 10.1007/978-3-319-05128-4
Springer Cham Heidelberg New York Dordrecht London

Printed on acid-free paper

Springer is part of Springer Science+Business Media (www.springer.com)

Preface

The laser belongs to one of the most fascinating fields in modern physics since its first experimental demonstration in 1960 by T. H. Maiman. The laser itself and its applications have fundamentally influenced many fields in modern physics as well as in many other sciences—some of which only become possible through the existence of the laser. The outstanding quantum-mechanical properties of laser radiation, for example its coherence and interaction with atoms or molecules, opened new fields of research, from spectroscopy in physics, chemistry and biology to information processing, materials science and general metrology, and to some of the probably most fascinating fields of physics: the laser allows us to create and study extreme states of matter such as Bose-Einstein condensates or degenerate Fermi gases. It opens a way to important investigations in quantum mechanics, has a big impact on solid-state physics and electronics by creating a need for more and more efficient light sources such as laser diodes and it made the demonstration and exploitation of the interesting field of non-linear optics possible. The laser will provide an enormous contribution also in the future, to discover gravitational waves, to create extremely hot and dense matter, for example for inertial fusion, and it opens the was to understand the fundamentals of physics at ultra-short time scales, which has become possible only owing to the existence of femto- and atto-second laser pulses.

This text book originates from a lecture in laser physics at the Karlsruhe School of Optics and Photonics at the Karlsruhe Institute of Technology (KIT), Karlsruhe, Germany, which has been given there since 2008. A main item in the conception of this text book was, to describe the fundamentals of lasers in a uniform and especially lab-oriented notation and formulation, as well as many currently well-known laser types, becoming more and more important in the future. It closes a gap between, for example, the measurable spectroscopic quantities and the whole theoretical description and modelling.

This text book contains not only the fundamentals and the context of laser physics in a mathematical and methodical approach important for university-level studies. It allows simultaneously, owing to its conception and its modern notation, to directly implement and use the learned matter in the practical lab work. It is presented in a format suitable for everybody, who wants to not only understand the fundamentals of lasers, but also use modern lasers or even develop and make laser setups. This text book tries to limit prerequisite knowledge and fundamental understanding to a minimum and is intended for students in physics, chemistry and mathematics after a bachelor degree, with the intention to create as much joy and interest as seen among the participants of the corresponding lecture.

This university text book describes in its first three chapters the fundamentals of lasers: light-matter interaction, the amplifying laser medium and the laser resonator. In the fourth chapter, pulse generation and related techniques are presented and investigated. The fifth chapter gives a closing overview on to different laser types gaining importance currently and in the future. It also serves as a set of examples, on which the theory learned in the first four chapters is applied and extended.

The author wishes to thank Prof. David H. Titterton (Dstl, UK) for proof reading the manuscript and offering valuable comments, and to Springer, here especially to Vera Spillner and Claus Ascheron for the extraordinary and friendly collaboration.

Saint Louis, France   Marc Eichhorn

Contents

Chapter 1
Quantum-Mechanical Fundamentals of Lasers

In this chapter we will investigate the basic quantum-mechanical effects and relations that allow the realization of a laser and determine the properties of laser operation. These are the fundamental processes of absorption, spontaneous emission and stimulated emission of light and their quantum-mechanical description.

1.1 Einstein Relations and Planck's Law

It was is the early years of quantum physics, when Planck found a theoretical description of the spectral distribution of the blackbody radiation. This radiation, which is emitted, e.g., from a small hole in the walls of a hohlraum (the blackbody) kept at a temperature T as shown in Fig. 1.1, shows a characteristic spectrum. Its spectral distribution and the peak of the emission intensity are only a function of the blackbody temperature. In Planck's derivation of this spectrum he assumed that electromagnetic radiation cannot be emitted or absorbed continuously, but only in fixed amounts of energy, the quanta, with a corresponding energy of

$$E = h\nu = \frac{hc}{\lambda}. \tag{1.1}$$

Today we know that these quanta are the photons of the electromagnetic field that can be described by their frequency ν or their wavelength λ.

Einstein also tried to find a derivation of this spectral distribution, starting from the fundamental interactions of absorption and emission between a quantum-mechanical system (atom, ion, molecule, electronic states in condensed matter for

Fig. 1.1 Measurement of the spectral distribution of the blackbody radiation emitted by a hohlraum at a temperature T

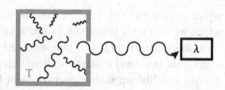

M. Eichhorn, *Laser Physics*, Graduate Texts in Physics,
DOI 10.1007/978-3-319-05128-4_1,
© Springer International Publishing Switzerland 2014

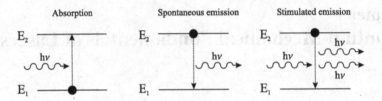

Fig. 1.2 Interactions between a two-level system and a photon according to Einstein

example) and a photon. According to Einstein, this can be described by three basic processes as shown in Fig. 1.2 for a simple two-level system. These processes are:

- The **absorption** of a photon of energy $h\nu = E_2 - E_1$, causing a transition from level $|1\rangle$ to level $|2\rangle$.
- The **spontaneous emission**, in which the system emits a photon of energy $h\nu$ by returning from level $|2\rangle$ to level $|1\rangle$. This is called spontaneous emission as the moment of emission (i.e. the phase ϕ of the radiation), the polarization $\vec{\epsilon}$ and the propagation direction, i.e. the direction of the wave vector $\vec{k}$, is random. Thus the spontaneous emission causes incoherent radiation and is responsible for the fluorescence of excited media.
- The **stimulated emission**, in which an incoming photon induces a resonant transition from the excited level $|2\rangle$ to level $|1\rangle$, emitting a second photon of energy $h\nu$. As photons are Bosons, i.e. they are allowed to be in the same quantum-mechanical state, and as stimulated emission is a resonant process, both photons are identical in all their properties. This effect, therefore, allows the amplification of light, the fundamental process of any laser.

The fundamental process that allows us to realize a laser is the stimulated emission process occurring in excited quantum-mechanical systems, giving rise to the possibility of photon amplification. It was the existence of the stimulated emission process that Einstein postulated in 1917 in order to derive the well-known **Planck's law** of the spectral energy density of electromagnetic radiation per volume $u(\nu, T)$ in the spectral range ν to $\nu + d\nu$ (or in wavelengths $u(\lambda, T)$ in the spectral range λ to $\lambda + d\lambda$),

$$u(\nu, T)d\nu = \frac{8\pi h \nu^3}{c^3} \frac{1}{e^{\frac{h\nu}{k_B T}} - 1} d\nu \tag{1.2}$$

$$u(\lambda, T)d\lambda = \frac{8\pi h c}{\lambda^5} \frac{1}{e^{\frac{hc}{\lambda k_B T}} - 1} d\lambda, \tag{1.3}$$

which is shown in Fig. 1.3. Therein, also the classical Rayleigh-Jeans law is shown, which we will need later to find some relations in Einstein's derivation.

In this derivation [1] Einstein assumed that an ensemble of $N = N_1 + N_2$ non-degenerate two-level systems with an energy difference $\Delta E = h\nu = E_2 - E_1$ is in thermal equilibrium with its environment kept at a temperature T. The absorption

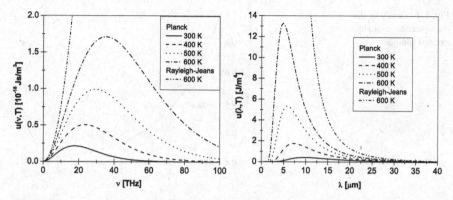

Fig. 1.3 Plot of the spectral energy density of electromagnetic radiation per volume as a function of frequency or wavelength for different temperatures

of the radiation then causes a transition rate from level $|1\rangle$ to level $|2\rangle$,

$$\left(\frac{dN_2}{dt}\right)_{abs} = -\left(\frac{dN_1}{dt}\right)_{abs} = B_{12}u(v,T)N_1, \tag{1.4}$$

that is proportional to the (unknown) radiation energy density $u(v,T)$ and the number of absorbers N_1 with a proportionality constant B_{12}. The stimulated emission of the radiation causes a transition from level $|2\rangle$ to level $|1\rangle$ with the rate

$$\left(\frac{dN_2}{dt}\right)_{stim} = -\left(\frac{dN_1}{dt}\right)_{stim} = -B_{21}u(v,T)N_2, \tag{1.5}$$

which is also proportional to the radiation density $u(v,T)$ and the number of emitters N_2 with a proportionality constant B_{21}. The spontaneous emission is only proportional to the number of the possible emitters N_2 and causes a rate

$$\left(\frac{dN_2}{dt}\right)_{spont} = -\left(\frac{dN_1}{dt}\right)_{spont} = -A_{21}N_2. \tag{1.6}$$

The proportionality constants B_{12}, B_{21} and A_{21} are called **Einstein coefficients**.

From Eq. (1.6) it can be deduced that in absence of other processes the population of level $|2\rangle$ decays exponentially with a time constant $\tau_{21} = A_{21}^{-1}$, called the natural lifetime of the level $|2\rangle$. Therefore, the evolution of the externally measurable fluorescence intensity $I(t) \propto \frac{dN_2}{dt}$ is given by

$$I(t) = I(0)e^{-\frac{t}{\tau_{21}}}. \tag{1.7}$$

This exponential decay is shown in Fig. 1.4.

In thermal equilibrium the populations of the levels $|1\rangle$ and $|2\rangle$ are constant, i.e.

$$\frac{dN_2}{dt} = \left(\frac{dN_2}{dt}\right)_{abs} + \left(\frac{dN_2}{dt}\right)_{stim} + \left(\frac{dN_2}{dt}\right)_{spont} = 0, \tag{1.8}$$

Fig. 1.4 Decay of the
fluorescence intensity of an
excited sample

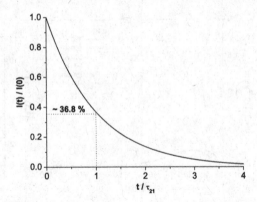

$$\frac{dN_1}{dt} = \left(\frac{dN_1}{dt}\right)_{abs} + \left(\frac{dN_1}{dt}\right)_{stim} + \left(\frac{dN_1}{dt}\right)_{spont} = 0, \qquad (1.9)$$

and their ratio can be described by a Boltzmann distribution resulting in

$$\frac{N_2}{N_1} = \frac{B_{12}u(v,T)}{A_{21} + B_{21}u(v,T)} \overset{!}{=} e^{-\frac{E_2-E_1}{k_BT}}. \qquad (1.10)$$

Therefore, the spectral energy density $u(v,T)$ has to have the form

$$u(v,T) = \frac{A_{21}}{B_{12}e^{\frac{hv}{k_BT}} - B_{21}}. \qquad (1.11)$$

In order to find the relations between the Einstein coefficients, two limits are investigated: In the high temperature limit $T \to \infty$ the spectral energy density diverges, forcing

$$B_{21} = B_{12}. \qquad (1.12)$$

This result is very important as it shows that absorption and stimulated emission are completely equivalent processes.

For the low photon energy limit $hv \ll k_BT$, $u(v,T)$ needs to be consistent with the classical **Rayleigh-Jeans law**

$$u_{RJ}(v,T) = \frac{8\pi v^2}{c^3}k_BT, \qquad (1.13)$$

which itself was proven experimentally and which can be deduced in the scope of classical Maxwellian electrodynamics, as it does not contain h. This comparison results in

$$A_{21} = \frac{8\pi hv^3}{c^3}B_{12}, \qquad (1.14)$$

stating that absorption and spontaneous emission are proportional to each other (**Kirchhoff's law**). Including both limits into Eq. (1.11) then gives Planck's law as in Eq. (1.2).

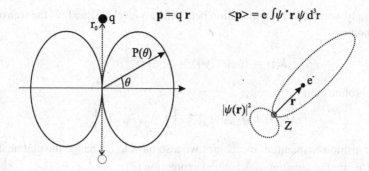

Fig. 1.5 Spatial radiation power of a classical dipole and dipole moment of an electron in quantum mechanics

1.2 Transition Probabilities and Matrix Elements

In this section we will derive the relations between the Einstein coefficients and the quantum-mechanical properties of a dipole transition [2].

1.2.1 Dipole Radiation and Spontaneous Emission

In classical electrodynamics, a dipole consisting of a charge q oscillating at a frequency $\omega = 2\pi \nu$ with a spatial amplitude $r_0 = |\vec{r}_0|$ possesses an electric dipole moment

$$\vec{p}(t) = q\vec{r}(t) = q\vec{r}_0 \sin \omega t. \tag{1.15}$$

As the oscillating dipole is an accelerated charge, it will give rise to a dipole radiation, see Fig. 1.5. The total radiated average power $\overline{P}$ can be derived in the scope of classical electrodynamics and results in **Larmor's formula**,

$$\overline{P} = \frac{2}{3} \frac{\overline{\vec{p}^2} \omega^4}{4\pi \epsilon_0 c^3}. \tag{1.16}$$

Therein,

$$\overline{f} = \frac{1}{T} \int_0^T f \, dt \tag{1.17}$$

is the time-average over one period $T = \frac{2\pi}{\omega}$, leading to

$$\overline{\vec{p}^2} = \frac{1}{2} q^2 |\vec{r}_0|^2. \tag{1.18}$$

In quantum mechanics, the average dipole moment of an electron with its elementary charge e, described by the wave function ψ, is given by

$$\langle \vec{p} \rangle = \langle \psi | e\vec{r} | \psi \rangle = \int \psi^* e\vec{r} \psi \, dV. \tag{1.19}$$

Accordingly, we define for a transition between two levels $|1\rangle$ and $|2\rangle$ the transition dipole moment

$$\vec{M}_{21} = \langle \psi_2 | e\vec{r} | \psi_1 \rangle = \int \psi_2^* e\vec{r} \psi_1 \, dV \tag{1.20}$$

and its absolute value

$$M_{21} = |\langle \psi_2 | e\vec{r} | \psi_1 \rangle| = \left| \int \psi_2^* e\vec{r} \psi_1 \, dV \right|. \tag{1.21}$$

In this transition to quantum mechanics we also have to exchange the classical average of $\overline{\vec{p}^2}$ by the quantum-mechanical expression [3]

$$\overline{\vec{p}^2} \rightarrow \frac{1}{2}(M_{21} + M_{12})^2 = 2M_{21}^2. \tag{1.22}$$

Inserting this in Eq. (1.16) results in an emitted power given by

$$\langle P_{21} \rangle = \frac{4}{3} \frac{\omega^4}{4\pi \epsilon_0 c^3} M_{21}^2, \tag{1.23}$$

with $\omega = (E_2 - E_1)/\hbar$. According to Eq. (1.6), the total average fluorescence power P_f emitted by N_2 excited levels corresponds to

$$P_f = h\nu A_{21} N_2 \overset{!}{=} \langle P_{21} \rangle N_2. \tag{1.24}$$

This comparison now allows us to deduce the explicit form of the Einstein coefficient A_{21} as

$$A_{21} = \frac{2}{3} \frac{e^2 \omega^3}{h \epsilon_0 c^3} |\langle \psi_2 | \vec{r} | \psi_1 \rangle|^2 = \frac{2}{3} \frac{e^2 \omega^3}{h \epsilon_0 c^3} \left| \int \psi_2^* \vec{r} \psi_1 \, dV \right|^2. \tag{1.25}$$

For an atom or molecule with many different levels for which the wave functions are known, the spontaneous emission rates A_{ji} may now be calculated for all possible transitions between the levels j and i, resulting in a matrix $A_{[j,i]}$. Therefore the M_{ji} in Eq. (1.21) are also called matrix elements.

The derivation above, as a result of Eqs. (1.16), (1.25), is valid only in the dipole approximation, i.e. as long as the wavelength of the emitted radiation is longer than the spatial dimension of the dipole $\lambda \gg r_0$. This is true for $\lambda > 1$ nm, and therefore, for all visible and infrared lasers.

From the ω^3 dependence in Eq. (1.25) it can also be concluded that the spontaneous emission increases dramatically for short wavelengths, resulting in very short lifetimes of the corresponding upper level. As will be shown in Chap. 2, this affords very high pump and laser intensities to saturate the optical transition, making the realization of deep-UV and X-ray lasers based on electronic transitions very difficult.

1.2.2 Stimulated Emission and Absorption

In contrast to the description of the spontaneous emission presented before, the stimulated emission or absorption of a photon by our two-level system is a quantum-mechanical process that involves the interaction between the system and the electromagnetic field. Therefore, it is necessary to make a short excursion into time-dependent quantum-mechanics and perturbation theory [4]:

Let $\mathbb{H}_0$ be the Hamiltonian for the unperturbed system, i.e. the system without the electromagnetic field, which is therefore described by the time-dependent Schrödinger equation

$$i\hbar\frac{\partial}{\partial t}\big|\psi^0(t)\big\rangle = \mathbb{H}_0\big|\psi^0(t)\big\rangle \tag{1.26}$$

for $t < t_0$ with $|\psi^0(t)\rangle$ being the state of the system before the perturbation occurs. After the application of the time-dependent perturbation $\mathbb{V}(t)$, which is assumed to be small compared with $\mathbb{H}_0$, the system will occupy the state $|\psi(t)\rangle$ and will evolve according to

$$i\hbar\frac{\partial}{\partial t}\big|\psi(t)\big\rangle = \big(\mathbb{H}_0 + \mathbb{V}(t)\big)\big|\psi(t)\big\rangle. \tag{1.27}$$

The time dependent perturbation theory allows to calculate the transition rates between different states. The exact derivation of the following formulas can be found, e.g., in [4]. Here we only quote the results that we need to investigate the stimulated emission and absorption processes. As the electromagnetic field of the incident photon can be treated as a periodic perturbation, we use the corresponding results of perturbation theory for a periodic perturbation oscillating at a frequency $\omega = 2\pi\nu$ of the form

$$\mathbb{V}(t) = \mathbb{F}e^{-i\omega t} + \mathbb{F}^\dagger e^{i\omega t}, \tag{1.28}$$

in which $\mathbb{F}$ is an operator defining the nature of the perturbation. Then the transition rate, i.e. the transition probability per unit time, for the transition from state j to state i can be calculated by **Fermi's golden rule**, resulting in

$$R_{ji} = \frac{2\pi}{\hbar}\Big(\delta(E_j - E_i - \hbar\omega)\big|\langle\psi_j|\mathbb{F}|\psi_i\rangle\big|^2 + \delta(E_j - E_i + \hbar\omega)\big|\langle\psi_j|\mathbb{F}^\dagger|\psi_i\rangle\big|^2\Big). \tag{1.29}$$

The two δ-functions describe the conservation of energy (or the resonance of the process) and the matrix elements of the perturbation $\mathbb{F}$ account for the strength of the transition. Owing to $\omega > 0$ it follows that for $E_j > E_i$ the first term describes the stimulated emission and for $E_j < E_i$ the absorption process is described by the second term.

For the stimulated emission from level $|2\rangle$ to level $|1\rangle$ or the absorption from level $|1\rangle$ to level $|2\rangle$ in our two-level system the perturbation is given by the electric field of the incoming photon

$$\vec{E}(t) = \vec{E}_0 e^{i\vec{k}\vec{r}} e^{-i\omega t}, \tag{1.30}$$

causing a transition rate [5]

$$R_{21} = \frac{\pi e^2}{2\hbar^2}\big|\langle\psi_2|\vec{E}_0\vec{r}e^{i\vec{k}\vec{r}}|\psi_1\rangle\big|^2\delta(\omega_0 - \omega) = \frac{\pi e^2}{2\hbar^2}\bigg|\int \psi_2^*\vec{E}_0\vec{r}e^{i\vec{k}\vec{r}}\psi_1 dV\bigg|^2\delta(\omega_0 - \omega). \tag{1.31}$$

Fig. 1.6 Local coordinate
frame for the calculation of
the average over all
polarizations

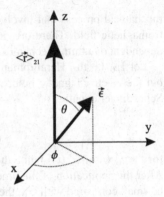

Therein $\vec{k}$ is the wave vector of the electromagnetic wave with $|\vec{k}| = \frac{2\pi}{\lambda}$ and $\omega_0 = \frac{E_2 - E_1}{\hbar}$ is the resonance frequency.

In the same dipole approximation as used for the spontaneous emission, $\lambda \gg r_0$, i.e. $\vec{k}\vec{r} \ll 1$, this rate can be approximated by taking $|e^{i\vec{k}\vec{r}}| \approx 1$ to give

$$R_{21} = \frac{\pi e^2}{2\hbar^2} E_0^2 |\vec{\epsilon} \langle \psi_2 | \vec{r} | \psi_1 \rangle|^2 \delta(\omega_0 - \omega) = \frac{\pi e^2}{2\hbar^2} E_0^2 \left| \vec{\epsilon} \int \psi_2^* \vec{r} \psi_1 dV \right|^2 \delta(\omega_0 - \omega),$$

$$(1.32)$$

with $\vec{\epsilon}$ describing the polarisation of the wave. It shows that the electric field needs to be applied in the same direction as the dipole orientation, in order to produce the maximum transition rate.

We will now simplify this relation for the case of a thermal radiation that is isotropically distributed in space. Therefore, Eq. (1.32) is averaged over all possible polarisation orientations $\vec{\epsilon}$, noting that $\langle \vec{r} \rangle_{21} = \int \psi_2^* \vec{r} \psi_1 dV$ is a constant vector after the integration has been performed. By defining a local frame with its z-axis aligned with $\langle \vec{r} \rangle_{21}$ as in Fig. 1.6, the average over all orientations of

$$\vec{\epsilon} = \begin{pmatrix} \sin\theta\cos\phi \\ \sin\theta\sin\phi \\ \cos\theta \end{pmatrix} \qquad (1.33)$$

is calculated in polar coordinates to give

$$\left\langle |\vec{\epsilon} \langle \vec{r} \rangle_{21}|^2 \right\rangle = \frac{1}{4\pi} \int_0^\pi \int_0^{2\pi} \cos^2\theta |\langle \vec{r} \rangle_{21}|^2 \sin\theta d\theta d\phi = \frac{1}{3} |\langle \vec{r} \rangle_{21}|^2. \qquad (1.34)$$

This results in the averaged transition rate

$$\langle R_{21} \rangle = \frac{\pi e^2}{6\hbar^2} E_0^2 |\langle \psi_2 | \vec{r} | \psi_1 \rangle|^2 \delta(\omega_0 - \omega) = \frac{\pi e^2}{6\hbar^2} E_0^2 \left| \int \psi_2^* \vec{r} \psi_1 dV \right|^2 \delta(\omega_0 - \omega),$$

$$(1.35)$$

which can be further simplified by introducing the spectral energy density of the electric field at resonance,

$$u(v) = \frac{1}{2}\epsilon_0 E_0^2 \delta(v_0 - v) = \pi \epsilon_0 E_0^2 \delta(\omega_0 - \omega), \tag{1.36}$$

using $\delta(ax) = \frac{1}{a}\delta(x)$. This yields

$$\langle R_{21} \rangle = \frac{2}{3}\frac{\pi^2 e^2}{3\epsilon_0 h^2}\left|\langle \psi_2|\vec{r}|\psi_1\rangle\right|^2 u(v) = \frac{2}{3}\frac{\pi^2 e^2}{3\epsilon_0 h^2}\left|\int \psi_2^* \vec{r} \psi_1 dV\right|^2 u(v). \tag{1.37}$$

The direct comparison with Eq. (1.5) results in the expression for the Einstein coefficient of stimulated emission,

$$B_{21} = \frac{2}{3}\frac{\pi^2 e^2}{\epsilon_0 h^2}\left|\langle \psi_2|\vec{r}|\psi_1\rangle\right|^2 = \frac{2}{3}\frac{\pi^2 e^2}{\epsilon_0 h^2}\left|\int \psi_2^* \vec{r} \psi_1 dV\right|^2. \tag{1.38}$$

It is, in contrast to A_{21}, independent of the transition frequency or wavelength, and does only depend on the quantum-mechanical properties of the transition enclosed in the matrix elements. By comparing this result with the Einstein coefficient of spontaneous emission A_{21} in Eq. (1.25), we again find the relation shown in Eqs. (1.12), (1.14).

1.3 Mode Structure of Space and the Origin of Spontaneous Emission

Spontaneous emission can be seen as a statistical process, i.e. each atom, ion or molecule decays independently by emitting a photon at a certain time in a single process whilst the observation of the fluorescence of an ensemble of many atoms, ions or molecules shows the well-known exponential decay law of Eq. (1.7). However, this statistical view cannot explain why a single atom "decides" to emit the photon at a certain time. In order to answer this question, we need to have a look into the mode structure of space, i.e. the structure of the allowed eigenmodes of electromagnetic radiation in vacuum, and the nature of the photons occupying these states.

1.3.1 Mode Density of the Vacuum and Optical Media

In order to determine the mode density of the vacuum and of optical transparent media with a refractive index $n > 1$, we first calculate the number of eigenmodes of a cubic hohlraum resonator of length a and volume a^3 up to the frequency ω. As we assume infinitely conductive walls the tangential components of the electric field must vanish on these walls. Therefore, the set of eigenmodes can be represented by

Fig. 1.7 Standing waves in a conductive hohlraum and representation of the eigenmodes in reciprocal space

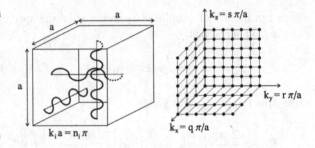

the standing waves inside the hohlraum as shown in Fig. 1.7 with the wave vectors given by

$$\vec{k} = \frac{\pi}{a}\begin{pmatrix} q \\ r \\ s \end{pmatrix} \quad \text{with } q, r, s \in \mathbb{Z}. \tag{1.39}$$

The corresponding electric field can be written as

$$\vec{E} = \vec{E}_0 \cos \omega t, \tag{1.40}$$

with the spatial components

$$\vec{E}_0 = \begin{pmatrix} E_{0x} & \cos\frac{\pi q}{a}x & \sin\frac{\pi r}{a}y & \sin\frac{\pi s}{a}z \\ E_{0y} & \sin\frac{\pi q}{a}x & \cos\frac{\pi r}{a}y & \sin\frac{\pi s}{a}z \\ E_{0z} & \sin\frac{\pi q}{a}x & \sin\frac{\pi r}{a}y & \cos\frac{\pi s}{a}z \end{pmatrix}, \tag{1.41}$$

the wave vector

$$|\vec{k}_{qrs}| = \frac{\pi}{a}\sqrt{q^2 + r^2 + s^2} \tag{1.42}$$

and the possible resonance frequencies

$$\omega_{qrs} = \frac{\pi c}{a}\sqrt{q^2 + r^2 + s^2}, \tag{1.43}$$

that result from the dispersion relation

$$\omega = c|\vec{k}| \tag{1.44}$$

of electromagnetic waves in vacuum.

In reciprocal space or k-space, where all eigenmodes are represented by a three-dimensional point lattice with a lattice constant $\frac{\pi}{a}$, Eq. (1.42) describes a sphere with radius $|\vec{k}| = \frac{\omega}{c}$. For high frequencies, i.e. large mode numbers $q^2 + r^2 + s^2 \gg 1$, the discrete lattice can be approximated by a homogeneous k-space density $\rho_k = (\frac{a}{2\pi})^3$, which takes into account that, e.g., $-q$ and q describe the same mode. This allows an easy calculation of the volume density of the number of modes in the hohlraum up to the frequency ν:

$$M(\nu) = 2a^{-3}\int \rho_k d^3k = 8\pi \int \frac{k^2 dk}{(2\pi)^3} = \frac{8\pi n^3}{c^3}\int \nu^2 d\nu = \frac{8\pi n^3 \nu^3}{3c^3}. \tag{1.45}$$

Therein the factor of 2 represents the two independent polarisations of the electric field and n the refractive index in the case of a hohlraum filled with an optical medium. This result is independent from the external dimension or orientation of the hohlraum. We can thus let $a \to \infty$ and find the spectral mode density of space as

$$\tilde{M}(\nu) = \frac{\partial M}{\partial \nu} = \frac{8\pi n^3 \nu^2}{c^3}. \tag{1.46}$$

One can recognize this spectral mode density in Planck's law, Eq. (1.2) and in the Rayleigh-Jeans law, Eq. (1.13), which simply states that each of these modes is excited with an energy of $k_B T$ in thermal equilibrium.

An alternative deduction of Planck's law, which is the one Planck used, starts from this spectral mode density $\tilde{M}(\nu)$, which is multiplied by the energy per photon $h\nu$ and by the number of thermally excited photons per mode

$$n(\nu, T) = \frac{1}{e^{\frac{h\nu}{k_B T}} - 1}, \tag{1.47}$$

to yield the spectral energy density in thermal equilibrium. $n(\nu, T)$ is given by the **Bose-Einstein distribution** as photons, spin 1 particles, are bosons.

1.3.2 Vacuum Fluctuations and Spontaneous Emission

We now know the spectral mode density of space and the fact that photons are bosons, which means especially that the number of photons in one quantum-mechanical state, i.e. in one mode, is not restricted. But what kind of state is a mode, i.e. what energy potential creates this state? In order to answer this question we have to look at the quantum structure of the electromagnetic field.

In classical electrodynamics [1], as we did for the determination of the spectral mode density, we can see a mode as a monochromatic wave, e.g. propagating along the x-axis and polarized along the z-axis with an electric field given by

$$\vec{E}(t) = \begin{pmatrix} 0 \\ 0 \\ p(t) \sin kx \end{pmatrix}, \tag{1.48}$$

with a temporal evolution described by $p(t)$. The corresponding magnetic field is given by the Maxwell-Equation $\vec{\nabla} \times \vec{E} = -\frac{\partial \vec{B}}{\partial t}$, resulting in

$$\vec{B}(t) = \begin{pmatrix} 0 \\ \frac{1}{c} q(t) \cos kx \\ 0 \end{pmatrix}, \tag{1.49}$$

with $q(t)$ being given by

$$\frac{dq(t)}{dt} = \omega p(t) \tag{1.50}$$

using Eq. (1.44). By inserting both fields into the Maxwell-Equation $\vec{\nabla} \times \vec{B} = \epsilon_0 \mu_0 \frac{\partial \vec{E}}{\partial t}$, we get

$$\frac{dp(t)}{dt} = -\omega q(t), \tag{1.51}$$

which we can combine with Eq. (1.50) to result in

$$\frac{d^2 q(t)}{dt^2} + \omega^2 q(t) = 0. \tag{1.52}$$

This is the equation of motion of a harmonic oscillator, that, in terms of classical mechanics, can be described by a Hamilton function, i.e. a total energy, of

$$H = \frac{1}{2}\omega\left(p^2 + q^2\right). \tag{1.53}$$

The quantization of the electromagnetic field can now be done by formally identifying this result with the quantum-mechanical harmonic oscillator of mass m described by the Hamiltonian

$$\mathbb{H}_{HO} = \frac{1}{2m}\mathbb{P}^2 + \frac{1}{2}m\omega^2\mathbb{Q}^2 = \hbar\omega\left(\mathbb{N} + \frac{1}{2}\right), \tag{1.54}$$

with $\mathbb{P}$ and $\mathbb{Q}$ being the momentum and position operators, respectively. For this formal identification the mass m is a free, non-used parameter that we can simply set to $m = \omega^{-1}$ and compare the result with the classical Eq. (1.53). The different operators in this Hamiltonian are given by

$$\mathbb{N} = \mathbb{A}^\dagger \mathbb{A} \tag{1.55}$$

$$\mathbb{A} = \frac{1}{\sqrt{2\hbar}}\left(\sqrt{m\omega}\mathbb{Q} + \frac{i}{\sqrt{m\omega}}\mathbb{P}\right) \tag{1.56}$$

$$\mathbb{A}^\dagger = \frac{1}{\sqrt{2\hbar}}\left(\sqrt{m\omega}\mathbb{Q} - \frac{i}{\sqrt{m\omega}}\mathbb{P}\right), \tag{1.57}$$

with $\mathbb{N}$ being the **occupation number operator** and $\mathbb{A}$ and $\mathbb{A}^\dagger$ the **annihilation** and **creation operators**, respectively, as they decrease or increase the quantum number n of the eigenstate $|n\rangle$ by one, i.e. $\mathbb{A}|n\rangle = \sqrt{n}|n-1\rangle$ and $\mathbb{A}^\dagger|n\rangle = \sqrt{n+1}|n+1\rangle$. For the quantum-mechanical oscillator, we know that the possible energy levels are evenly spaced with a separation of $\hbar\omega$ as shown in Fig. 1.8 and that the energy eigenvalues are given by

$$\mathbb{H}_{HO}|n\rangle = E_n|n\rangle \quad \text{with } E_n = \hbar\omega\left(n + \frac{1}{2}\right). \tag{1.58}$$

As a result, we can state that a photon mode consisting of n photons can be seen as a harmonic quantum oscillator occupying the state with quantum number n. The linear relationship between the number of photons n and the total energy E_n of the mode is as expected, since adding a photon just should increase the energy of that mode by $\hbar\omega$. However, as it is well known from the quantum oscillator,

Fig. 1.8 Potential and
corresponding eigenstates of
the quantum-mechanical
harmonic oscillator

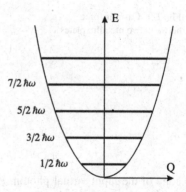

the lowest state $|0\rangle$ has a finite energy of $\frac{1}{2}\hbar\omega$, and since usually the quantum-mechanical system is found in its ground state at absolute zero, i.e. at $T = 0$ K, this energy is called **zero-point energy**. We thus found that in quantum electrodynamics the vacuum is not just an empty space. Electrodynamically, it can be described by an infinite set of modes that are all in their ground state, i.e. unoccupied. However, from **Heisenberg's uncertainty relation**

$$\Delta E \Delta t \geq \frac{\hbar}{2}, \tag{1.59}$$

we can deduce that it is quantum dynamically allowed that a photon of energy $\hbar\omega$ may be spontaneously created "out of nothing' for a short period of time $\Delta t \leq \frac{1}{2\omega}$, so that the necessary energy $\hbar\omega$ is within the theoretically allowed uncertainty

$$\Delta E \geq \frac{\hbar}{2\Delta t} \geq \hbar\omega. \tag{1.60}$$

Without going into detail here, we just state the result that these "virtual" photons do exist and are called **vacuum fluctuations**. In a classical picture this explanation is not possible as the time $\Delta t \leq \frac{1}{2\omega}$ does not allow a full oscillation to occur.

An experimental proof of the vacuum fluctuations, e.g., the **Casimir force**, an additional attractive force component between two uncharged parallel metallic plates placed close to each other. It can be explained by the fact that between the plates only those vacuum fluctuations occur that are consistent with the allowed standing modes, see illustration in Fig. 1.9, whilst in the outer space all vacuum fluctuations occur. Thus an external pressure is created that pushes the two plates together.

Now that we know that the physical vacuum is no "quiet" space, but that vacuum fluctuations occur, we can understand the spontaneous emission in a much more fundamental way. By using the quantum nature of the electromagnetic field, being present in the form of modes that are occupied by a number of photons, we can separate this electromagnetic field into a real part consisting of real photons, and a virtual part, consisting of the virtual photons of the vacuum fluctuations. Therefore, we can describe the spontaneous emission process as a stimulated emission process, triggered by a virtual photon from the vacuum fluctuations. As in standard stimulated emission triggered from a real photon, the emitted photon here is an exact

Fig. 1.9 Casimir force
between two metallic plates

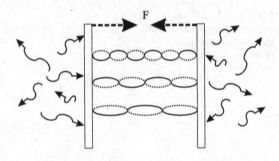

copy of the input virtual photon. However, no real photon amplification occurs as
the virtual photon has to disappear after the process to obey Eq. (1.59). The statistical behaviour of the vacuum fluctuations, i.e. the random nature of the time of
creation of the virtual photon as well as the mode in which it occurs, is therefore,
transferred on to the whole emission process, explaining the statistical nature of the
spontaneous emission. As a result of this more fundamental view, we find that any
state of a quantum-mechanical system that couples to the electromagnetic field, and
which is not the ground state of the system, will show spontaneous emission towards
the energetically lower lying states.

1.4 Cross Sections and Broadening of Spectral Lines

We will introduce in this section the spectroscopic properties that describe a laser
medium as well as the line broadening mechanisms that influence the spectral behaviour and the efficiency of a laser. This allows us to quantify the different properties of the optical transitions in a laser medium and results in a general mathematical
description of lasers that will be presented in Chap. 2.

1.4.1 Cross Sections of Absorption and Emission

When an electromagnetic wave propagates in an absorbing medium along the z-
axis, its intensity $I(z)$ will be attenuated during propagation. In this process each
frequency or wavelength component of the radiation may suffer from a different
absorption strength. Therefore, we introduce the spectral intensity $\tilde{I}(z, \lambda)$, which is
defined by

$$I(z) = \int \tilde{I}(z, \lambda)d\lambda. \tag{1.61}$$

By passing an infinitesimal propagation distance dz each wavelength component is
attenuated proportional to the incident spectral intensity according to

$$\frac{d\tilde{I}(z, \lambda)}{dz} = -\alpha(\lambda)\tilde{I}(z, \lambda), \tag{1.62}$$

as shown in Fig. 1.10.

Fig. 1.10 Absorption of light in a medium and geometrical interpretation of the cross section in an absorption process of particles

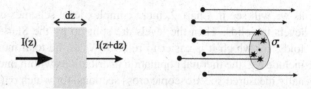

Integrating this equation under the assumption of a spatially constant absorption coefficient $\alpha(\lambda)$ leads to the **Lambert-Beer law** of absorption

$$\tilde{I}(z, \lambda) = \tilde{I}(0, \lambda)e^{-\alpha(\lambda)z}. \tag{1.63}$$

In the important case in which the absorption is caused by an optical transition from a lower state $|1\rangle$ to an upper state $|2\rangle$, as described by Eq. (1.4), the absorption coefficient is proportional to the number density N_1 of atoms, ions, molecules or other laser species in state $|1\rangle$ and can be written as

$$\alpha(\lambda) = \sigma_a(\lambda)N_1. \tag{1.64}$$

The proportionality constant $\sigma_a(\lambda)$ is the **absorption cross section**. It has a dimension of an area and can be interpreted as an effective "cross-sectional area" attached to, e.g., an atom that absorbes the incident photons as shown in Fig. 1.10. However, depending on the strength of the transition it can vary for different transitions in one atom and one should not confuse it with the geometrical size of the atom itself.

In the same way also the stimulated emission can be described as an amplification of the incident light according to

$$\frac{d\tilde{I}(z, \lambda)}{dz} = \gamma(\lambda)\tilde{I}(z, \lambda), \tag{1.65}$$

resulting in

$$\tilde{I}(z, \lambda) = \tilde{I}(0, \lambda)e^{\gamma(\lambda)z}. \tag{1.66}$$

In analogy to the absorption coefficient, the emission coefficient is proportional to the number density N_2 of atoms in the upper state

$$\gamma(\lambda) = \sigma_e(\lambda)N_2, \tag{1.67}$$

for which the proportionality constant $\sigma_e(\lambda)$ is the **emission cross section**. Taking both processes together leads to the total evolution of the spectral intensity given by

$$\tilde{I}(z, \lambda) = \tilde{I}(0, \lambda)e^{(\sigma_e(\lambda)N_2 - \sigma_a(\lambda)N_1)z} \tag{1.68}$$

In the special case of the two-level system in Fig. 1.2, where N_i denote the population densities of the two levels $|i\rangle$, it follows from Eq. (1.12) that the emission and absorption cross sections related to the intrinsic transition are equal, i.e. $\sigma_e(\lambda) = \sigma_a(\lambda)$. They are, therefore, called intrinsic cross sections. However, as we will see in Chap. 2, more complex level schemes exist, especially for ionic levels in solids. Then the levels are split up by the Stark effect and create manifolds, for which it is easier to refer to N_i as the total manifold population and to include e.g. the thermal population distributions within one manifold into the externally measured spectroscopic cross sections, for which $\sigma_e(\lambda)$ and $\sigma_a(\lambda)$ are usually different. This will be explained in more detail in Chap. 2.

To connect this spectroscopic description to the Einstein coefficients, the emitted spectral power density per volume resulting from stimulated emission in the medium of volume $dV = dA\,dz$ is investigated. Therefore, we use the description of Eq. (1.5) and assume that each emission process emits the energy of one photon $h\nu$ into the propagating mode. The photon energies itself are taken to be distributed around $h\nu_0 = E_2 - E_1$, given by a normalized distribution $\rho_f(\nu)$ describing this fluorescence, with

$$\int \rho_f(\nu)d\nu = 1. \tag{1.69}$$

$\rho_f(\nu)$ thus determines which frequencies are amplified by the stimulated emission. Then the emitted spectral power density per volume is given by

$$\frac{\partial \tilde{P}}{\partial V} = -h\nu\rho_f(\nu)\frac{\partial N_2}{\partial t} = h\nu\rho_f(\nu)B_{21}u(\nu)N_2. \tag{1.70}$$

The spectroscopic view on the other hand results in

$$\frac{\partial \tilde{P}}{\partial V} = \frac{\partial \tilde{I}}{\partial z} = \gamma(\nu)\tilde{I}(\nu) = N_2\sigma_e(\nu)\frac{c}{n}u(\nu) \tag{1.71}$$

Therein, it was assumed that $\tilde{I}(\nu)$ describes a collimated homogeneous beam, which is related to the energy density by $\tilde{I}(\nu) = \frac{c}{n}u(\nu)$ and describes that the energy "flows" with the velocity of light $\frac{c}{n}$ in a medium with a refractive index n.

Comparing Eq. (1.70) and Eq. (1.71) results in the relation

$$\sigma_e(\nu) = \frac{h\nu n}{c}B_{21}\rho_f(\nu). \tag{1.72}$$

By using the relation for the Einstein coefficients Eq. (1.14) and Eq. (1.12), which in a medium with a refractive index n changes owing to the changed mode density of Eq. (1.46) to

$$A_{21} = \frac{8\pi h\nu^3 n^3}{c^3}B_{21}, \tag{1.73}$$

the important relation

$$\sigma_e(\nu) = \frac{c^2}{8\pi n^2\nu^2\tau_{21,sp}}\rho_f(\nu) \tag{1.74}$$

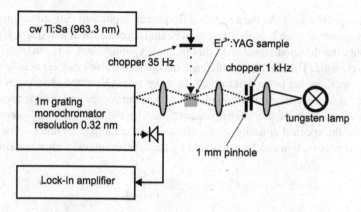

Fig. 1.11 Measurement setup for the determination of absorption and emission cross sections, e.g. of an Er^{3+}:YAG sample

results. Therein, it has been explicitly written that the A_{21} coefficient is related to the spontaneous decay with a decay time $\tau_{21,sp}$ as

$$A_{21} = \frac{1}{\tau_{21,sp}}. \tag{1.75}$$

Thus, the spectral distribution $\rho_f(\nu)$ of the light emitted from the volume dV is closely linked to the spectroscopic emission cross section $\sigma_e(\nu)$. Finally, by exploiting the normalization of $\rho_f(\nu)$ a relation between the upper state lifetime and the integral emission cross section can be deduced,

$$\frac{1}{\tau_{21,sp}} = \frac{8\pi n^2}{c^2} \int \sigma_e(\nu)\nu^2 d\nu = 8\pi n^2 c \int \frac{\sigma_e(\lambda)}{\lambda^4} d\lambda. \tag{1.76}$$

In the last step we used $|d\nu| = \frac{c}{\lambda^2}|d\lambda|$.

Equation (1.76) is called **Füchtbauer-Ladenburg relation**, for which it has to be noted that it is also valid for spectroscopic cross sections and that λ always refers to vacuum wavelengths. It allows the calculation of the **spontaneous emission lifetime** $\tau_{21,sp}$, also called **radiative lifetime**, from measured spectra, or, in the reverse sense, the calibration of measured spectral intensities $\tilde{I}(\lambda)$ to deduce the absolute values of the emission cross section $\sigma_e(\lambda)$.

For this application, the spectral fluorescence is recorded from an excited sample and the emission cross section is then calculated by

$$\sigma_e(\lambda) = \frac{\lambda^4 \tilde{I}(\lambda)}{8\pi n^2 c \tau_{21,sp} \int \tilde{I}(\lambda) d\lambda}. \tag{1.77}$$

A scheme of this measurement setup is shown in Fig. 1.11. An Er^{3+}:YAG sample is excited by the emission from a Ti:sapphire laser and its fluorescence is recorded by a 1 − m spectrometer. As the number of fluorescence photons that are captured by the spectrometer is often low, the excitation beam is modulated at a frequency f_{mod}, so that only the detected signal with this frequency component is recorded using a lock-in technique. This enables an increase in signal-to-noise ratio, especially when low cross sections are to be determined. In order to measure the absorption cross section, a broad-spectrum tungsten lamp is used as the source and the spectrometer records the spectrum of the intensity transmitted by the sample. After correcting the data for the spectral emission characteristics of the lamp (cf. Planck's law), the absorption cross section can be calculated by the relative intensity change using

$$\sigma_a(\lambda) = -\frac{1}{LN_1} \ln \frac{I_t(\lambda)}{I_{0,t}(\lambda)}. \tag{1.78}$$

Therein, N_1 equals the Er^{3+} ion density as the excitation power from the lamp is chosen to be low enough in order not to bleach the ground state, L is the length of the sample, $I_t(\lambda)$ the transmitted intensity signal and $I_{0,t}(\lambda)$ the reference recorded with an undoped sample in place.

1.4.2 Natural Line Width and Broadening of Spectral Lines

As a result of the Heisenberg uncertainty principle, Eq. (1.59), a transition between two quantum-mechanical levels cannot be infinitely sharp when the corresponding upper level has a finite lifetime τ, i.e. the corresponding cross sections $\sigma_e(\nu)$ and $\sigma_a(\nu)$ as well as the fluorescence distribution $\rho_f(\nu)$ discussed previously are no δ-functions. As shown in Sect. 1.3.2, each level above the ground state will at least have its natural lifetime that is determined by the spontaneous emission, and thus, by the vacuum fluctuations. Therefore, any optical transition will show a minimum line width, called the **natural line width** of the transition and the spectral line can be represented by its line form function $g(\nu)$, which is identical with the fluorescence distribution $\rho_f(\nu)$.

The fact that a laser medium usually consists of many identical absorption and emission systems, i.e. the atoms, ions or molecules, divides the interaction between them and the electromagnetic field into two cases that define the two different line broadening mechanisms:

- **Homogeneous line broadening**: In this case all systems show the same transition frequency ν_0, line width $\Delta\nu$ and form function $g(\nu)$. Therefore, they all contribute to the emission or absorption of a photon of energy $h\nu$ in the same way, i.e. with the same probability. They can be described by processes that reduce the upper level lifetime in a homogeneous way for all the systems, thus causing an equal broadening of all systems around the same resonance frequency ν_0, e.g. spontaneous emission (**natural line width**), lattice vibrations (phonons) of the crystal matrix in solid-state lasers causing **multi-phonon relaxation**, atomic collisions in gas lasers causing collisional relaxation (**pressure broadening**).

- **Inhomogeneous line broadening**: In this case, the transition frequency of the different systems varies, resulting in different interaction probabilities between a photon of energy $h\nu$ and the different systems. This inhomogeneous distribution of the resonance frequencies over the different systems may be temporally constant as e.g. in ion-doped amorphous solids as fibers, in which the **Stark effect** shifts the energy levels by a fixed amount for a given ion but varies from site to site in the glass matrix. It can also be time dependent for a given system as e.g. the Doppler-shift in gas lasers, which depends on the local velocity of an atom or molecule (**Doppler broadening**). Thus, the atom itself changes its resonance frequency over time resulting from its collisions and the corresponding changes in velocity and direction. However, for the ensemble, a constant effective average results from the Maxwell distribution.

Homogeneous Broadening

To find the line width function of a homogeneously broadened line $g(\nu)$ we will investigate a classical example [6]. The spontaneous emission can be explained by a suddenly emitted electric field oscillating at the resonance $\hbar\omega_0 = h\nu_0 = E_2 - E_1$ with an exponentially decaying amplitude with a time constant of 2τ (the time constant τ corresponds to the intensity $I \propto E^2$),

$$
E(t) = \begin{cases} 0 & \text{for } t \le 0 \\ E_0 e^{-\frac{t}{2\tau}} \cos \omega_0 t & \text{for } t > 0 \end{cases} = \begin{cases} 0 & \text{for } t \le 0 \\ \frac{E_0}{2} e^{-\frac{t}{2\tau}} \left(e^{i\omega_0 t} + e^{-i\omega_0 t} \right) & \text{for } t > 0 \end{cases}.
$$

(1.79)

The spectral components of this electric field are given by its Fourier transformation

$$
E(\nu) = \int_{-\infty}^{\infty} E(t) e^{-i2\pi \nu t} dt = i \frac{E_0}{4\pi} \left(\frac{1}{\nu_0 - \nu + \frac{i}{4\pi\tau}} - \frac{1}{\nu_0 + \nu - \frac{i}{4\pi\tau}} \right),
$$

(1.80)

resulting in the spectral intensity given by a **Lorentzian function**

$$
\tilde{I}(\nu) = I g(\nu) = \sqrt{\frac{\epsilon_0}{\mu_0}} |E(\nu)|^2 = I \frac{2}{\pi} \frac{\Delta\nu}{4(\nu - \nu_0)^2 + (\Delta\nu)^2}.
$$

(1.81)

Therein, $\Delta\nu = \frac{1}{2\pi\tau}$ is the natural line width, given as the full width at half maximum (FWHM) of the line.

Inhomogeneous Broadening

The most important broadening process in gas lasers is Doppler broadening, which will be described here as an example for inhomogeneous broadening. In a gas the different atoms or ions will show a kinetic velocity distribution with respect to one propagation direction, e.g. the z-axis, that is given by the normalized Maxwell distribution [6]

$$P(v_z) = \sqrt{\frac{m}{2\pi k_B T}} e^{-\frac{m v_z^2}{2 k_B T}}. \tag{1.82}$$

Therein, m is the mass of the atom, v_z its velocity component along the z-axis and T the kinetic gas temperature. While the atoms itself still have the same resonance frequency v_0 in their local rest frame, an external observer looking along the z-axis, however, will see the Doppler shifted frequency

$$v = v_0 \left(1 + \frac{v_z}{c} \right), \tag{1.83}$$

which is taken here in the non-relativistic limit. As the velocity distribution directly determines the probability of that velocity component in the gas, the line form function can be directly deduced as

$$g(v) = \frac{2\sqrt{\pi \ln 2}}{\pi \Delta v_D} e^{-(2\frac{v-v_0}{\Delta v_D})^2 \ln 2}, \tag{1.84}$$

with a Doppler line width of

$$\Delta v_D = v_0 \sqrt{\frac{8 k_B T \ln 2}{m c^2}}. \tag{1.85}$$

In contrast to the homogeneous broadening, the form function of the inhomogenous Doppler broadening is a **Gaussian function**.

Simultaneous Broadening Processes

In the case of different homogeneous broadening processes acting simultaneously, e.g. spontaneous emission and multi-phonon relaxation in a solid-state laser, each process contributes to the total decay of the upper level and can be described by its own lifetime or decay constant. In this example, they are the spontaneous lifetime τ_{sp} and the non-radiative relaxation lifetime τ_r, that both contribute to the decay of the upper level as

$$\frac{dN_2}{dt} = -\frac{N_2}{\tau_{sp}} - \frac{N_2}{\tau_r} = -\frac{N_2}{\tau}. \tag{1.86}$$

Therefore, for different homogeneous broadening processes the different lifetimes add inversely to the total lifetime τ like the parallel connection of resistors,

$$\frac{1}{\tau} = \sum_i \frac{1}{\tau_i}, \tag{1.87}$$

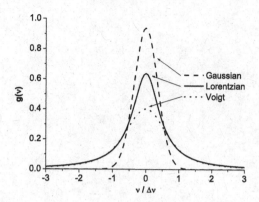

Fig. 1.12 Fundamental Lorentzian and Gaussian line shape functions and their convolution, the Voigt profile

and the corresponding line widths add directly like the parallel connection of capacitors,

$$\Delta v = \sum_i \Delta v_i. \tag{1.88}$$

The line shape function $g(v)$ of the combined line shape is again a Lorentzian function.

For the case of two inhomogeneous processes with line form functions $g_1(v)$ and $g_2(v)$, or a mixing between a homogeneous and an inhomogeneous process, the total line form function is in general given by the convolution of the line form functions of the different processes [6],

$$g(v) = \int_{-\infty}^{\infty} g_1(v')g_2(v - v')dv'. \tag{1.89}$$

Two Gaussian line shapes thus result in a new Gaussian line shape with the line width

$$\Delta v = \sqrt{\Delta v_1^2 + \Delta v_2^2}, \tag{1.90}$$

while for a mixing between a Lorentzian and a Gaussian line form function the convolution cannot be solved analytically and results in the **Voigt profile**. All three line shapes are shown for comparison in Fig. 1.12.

References

1. H. Haken, H.C. Wolf, *Atom- und Quantenphysik*, 7th edn. (Springer, Berlin, 2000), p. 59ff
2. W. Demtröder, *Experimentalphysik 3*, 2nd edn. (Springer, Berlin, 2000), p. 222f
3. W. Weizel, *Lehrbuch der theoretischen Physik*, vol. 2 (Springer, Berlin, 1958), p. 908
4. F. Schwabl, *Quanten-Mechanik* (Springer, Berlin, 1988), p. 262
5. S. Fluegge, *Rechenmethoden der Quantentheorie* (Springer, Berlin, 1993)
6. F.K. Kneubühl, M.W. Sigrist, *Laser* (Teubner, Stuttgart, 1999)

Chapter 2
The Laser Principle

After the investigation of the fundamental quantum-mechanical properties of absorption and emission in quantum systems, we are now able to realize and to describe a laser. The laser principle itself is already given in its arconym: **LASER** stands for **L**ight **A**mplification by **S**timulated **E**mission of **R**adiation. Lasers are thus based on the stimulated emission process that was postulated by Einstein in 1917. However, it took 43 years until the first laser, a ruby (Cr^{3+}:Al_2O_3) solid-state laser, was demonstrated by Maiman in 1960.

According to Eq. (1.66), the stimulated emission causes amplification of light and we can therefore, create a light oscillator by introducing feedback to the amplifying medium, as used in electronic oscillators. However, as shown by Eq. (1.12), the light may also be absorbed by this medium with an identical strength, making it more difficult to get an efficient amplification. In the next section we will, therefore, investigate the fundamental relations necessary for laser operation.

2.1 Population Inversion and Feedback

By combining Eq. (1.63) with Eq. (1.66), we find for the total amplification of an incident spectral intensity $\tilde{I}(0, \lambda)$

$$\tilde{I}(z, \lambda) = \tilde{I}(0, \lambda)e^{(\gamma(\lambda) - \alpha(\lambda))z} \tag{2.1}$$

and thus an effective gain $G(z, \lambda)$ of

$$G(z, \lambda) = \frac{\tilde{I}(z, \lambda)}{\tilde{I}(0, \lambda)} = e^{(\sigma_e(\lambda)N_2 - \sigma_a(\lambda)N_1)z}. \tag{2.2}$$

In order to amplify the incoming light, we need $G > 1$, thus

$$\sigma_e(\lambda)N_2 > \sigma_a(\lambda)N_1 \quad \text{or} \quad N_2 > \frac{\sigma_a(\lambda)}{\sigma_e(\lambda)}N_1. \tag{2.3}$$

This is called effective population inversion and states that there have to be more "systems" in the upper level that cause emission processes than there are "systems"

M. Eichhorn, *Laser Physics*, Graduate Texts in Physics,
DOI 10.1007/978-3-319-05128-4_2,
© Springer International Publishing Switzerland 2014

in the lower level causing absorption processes. For a single atomic line ($\sigma_a(\lambda) = \sigma_e(\lambda)$) this simplifies to

$$N_2 > N_1, \tag{2.4}$$

meaning that refering to the population densities of the two levels involved in the transition, there need to be more systems in the upper state than in the lower state. However, in thermal equilibrium the population density relation between the two levels is directly given by the Boltzmann distribution

$$\frac{N_2}{N_1} = e^{-\frac{h\nu_{21}}{k_B T}}. \tag{2.5}$$

A population inversion, can therefore, never be reached in thermal equilibrium, i.e. in a static process, and thus, we need a dynamic process that constantly achieves the non-equilibrium state described by Eq. (2.4).

2.1.1 The Two-Level System

In this section we will investigate if the laser condition of population inversion can be reached in a simple **two-level scheme** as shown in Fig. 1.2 by optical pumping, i.e. by providing an incident pump radiation that is absorbed by the two-level system. From Eq. (1.8) it follows in the steady-state that

$$\frac{dN_2}{dt} = B_{12}u(\nu)(N_1 - N_2) - A_{21}N_2 \overset{!}{=} 0 \tag{2.6}$$

$$\Rightarrow \quad \frac{N_2}{N_1} = \frac{B_{12}u(\nu)}{B_{12}u(\nu) + A_{21}} < 1 \quad \forall u(\nu) > 0. \tag{2.7}$$

This shows that by pumping a two-level system a steady-state population inversion cannot be obtained, and therefore, a two-level system does not lead to a laser process. Even in the case of infinite pump intensity, corresponding to $u(\nu) \to \infty$, we only approach an equilibrium between the two populations $N_2 \to N_1$. Although this negative result was shown here for optical pumping, it can be quantum mechanically proved that any other pumping mechanism (chemical reaction, electric discharge for example) would give the same result for a true two level system.

2.1.2 Three- and Four-Level Systems

The problem with the two-level system is inherent to quantum mechanics and is expressed by the symmetry of the absorption and emission process in Eq. (1.12). This symmetry is only broken by the spontaneous emission in the two-level system, resulting in $\frac{N_2}{N_1} < 1$. In order to circumvent this problem and to create an inversion, we need to avoid the stimulated emission into the pump wave. This is fulfilled if the stimulated emission towards the lower level occurs at a different wavelength from that of the (pump) absorption, and therefore, at minimum three levels are needed as shown in Fig. 2.1.

Fig. 2.1 Three level system
and corresponding transitions

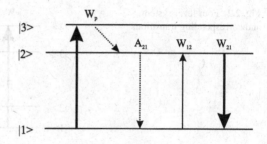

The Three-Level Laser

In this **three-level scheme** the pump radiation $u_p(\nu_p)$ pumps the level $|3\rangle$ with a
rate

$$\left(\frac{dN_3}{dt}\right)_{pump} = B_{13}u_p(\nu_p)N_1 = W_pN_1. \tag{2.8}$$

Therein, we define W_p as the pump rate of the transition $|1\rangle \rightarrow |3\rangle$. We now assume
that a very fast relaxation occurs from level $|3\rangle$ to level $|2\rangle$, e.g. induced by phonons
in a solid host crystal. This relaxation needs to occur at a rate much faster than the
pumping of level $|3\rangle$. Then no efficient stimulated back-emission into the pump can
occur, as the absolute population of level $|3\rangle$ stays low, i.e. $N_3 \approx 0$. We can then
assume that the pump rate of level $|2\rangle$, i.e. the change in the population of level $|2\rangle$
per unit time as a result of the pumping process, equals W_pN_1. Including emission
and absorption the steady-state rate equations for the two laser levels read

$$\frac{dN_2}{dt} = W_pN_1 + W_{12}N_1 - W_{21}N_2 - A_{21}N_2 \tag{2.9}$$

$$= W_pN_1 - W_{21}(N_2 - N_1) - A_{21}N_2 \overset{!}{=} 0 \tag{2.10}$$

$$N_2 + N_1 \approx N = \text{const}, \tag{2.11}$$

which results in

$$\frac{N_2}{N_1} = \frac{W_p + W_{21}}{A_{21} + W_{21}} > 1 \quad \text{for } W_p > \frac{1}{\tau_2}. \tag{2.12}$$

Therein, $\tau_2 = A_{21}^{-1}$ is the upper level lifetime. Thus, a population inversion can be
achieved in a three-level system. The pump rate necessary to achieve the inversion,
i.e. the pump intensity, depends on the spontaneous emission rate and has to compensate for the spontaneous losses to restore the broken symmetry. The pump rate
needs to be stronger for media with strong spontaneous emission, i.e. for media with
a short upper level lifetime.

However, this three-level scheme is still very inefficient, as at least 50 % of the
total population (assuming $N_3 \approx 0$) needs to be pumped into the upper laser level
$|2\rangle$ to provide $G > 1$. Therefore, only keeping the inversion at that level needs the
minimum pump rate of $A_{21}N_2$, and thus results in a high **laser threshold**. The laser
threshold itself is defined as the pump power necessary to initiate laser oscillation.

Fig. 2.2 Four level system and corresponding transitions

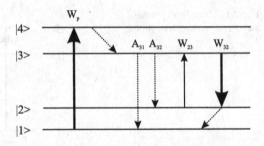

In order to keep this threshold as low as possible, we have to reduce the need for a high upper level population N_2 to reach inversion. This can be done by using the same idea as in the transition from the two-level to the three-level system: We have to reduce the lower laser level population by a relaxation process, which leads us to the **four level scheme** shown in Fig. 2.2.

The Four-Level Laser

In such a four-level system the pump rate W_p can be defined by

$$\left(\frac{dN_4}{dt}\right)_{pump} = B_{14}u_p(\nu_p)N_1 = W_pN_1. \tag{2.13}$$

As in the three-level system, we can assume that a very fast relaxation occurs from level $|4\rangle$ to level $|3\rangle$ at a rate much faster than the pumping of level $|4\rangle$. In addition, the lower laser level $|2\rangle$ is now no longer the ground-state of the system and we assume a second fast relaxation from level $|2\rangle$ to the ground-state $|1\rangle$. We also assume that the energy separation between the levels $|2\rangle$ and $|1\rangle$ is large enough so that level $|2\rangle$ is not thermally populated from level $|1\rangle$. Then we can state $N_4 \approx 0$ and $N_2 \approx 0$, which leads to the rate equation

$$\frac{dN_3}{dt} = W_pN_1 + W_{23}N_2 - W_{32}N_3 - (A_{32} + A_{31})N_3 \tag{2.14}$$

$$\approx W_pN_1 - W_{32}N_3 - (A_{32} + A_{31})N_3 \overset{!}{=} 0 \tag{2.15}$$

$$N_3 + N_1 \approx N = \text{const} \tag{2.16}$$

and thus to

$$N_3 = \frac{W_p}{W_p + W_{32} + A_{32} + A_{31}}N > 0 \quad \forall W_p > 0. \tag{2.17}$$

As the lower level population is $N_2 \approx 0$, the laser radiation on the transition $|3\rangle \to |2\rangle$ does not suffer from reabsorption and any population of level $|3\rangle$ causes a population inversion to occur. This four-level laser, will therefore, show the lowest possible threshold.

The Quasi-Three-Level Laser

In reality the approximations made for the three- and four-level laser are not always fulfilled, resulting in a net lower level population N_2 and also in a reduced but existing back-emission into the pump light. This often comes from the low separation between the levels $|1\rangle$ and $|2\rangle$ as well as $|4\rangle$ and $|3\rangle$, especially in the case of solid-state lasers. As this laser scheme can be seen to be in between the three- and the four-level laser, it is called **quasi-three-level scheme**. Sometimes also **quasi-four-level scheme** can be found in the literature. In the following we will deduce some important relations for quasi-three-level lasers, as well as motivate this description.

The spectroscopic rate equation description of a laser medium, as it is necessary to completely describe the quasi-three-level laser, can be viewed as the most general description, and will therefore, be investigated in detail in Sect. 2.2. Here we follow the simple description to compare the quasi-three-level system with the ones described before. We assume again that a very fast relaxation occurs from level $|4\rangle$ to level $|3\rangle$ at a rate much faster than the pumping of level $|4\rangle$. However, the lower laser level $|2\rangle$ is now so close to the ground-state $|1\rangle$ that it will be thermally populated. As this thermalization is much faster than the other processes discussed here, we can assume that the population density of $|2\rangle$ is a constant fraction $f > 0$ of the population density of $|1\rangle$,

$$N_2 = f N_1. \tag{2.18}$$

This leads to the rate equation

$$\frac{dN_3}{dt} = W_p N_1 + W_{23} N_2 - W_{32} N_3 - (A_{32} + A_{31}) N_3 \tag{2.19}$$

$$= \frac{W_p}{f} N_2 + W_{32} N_2 - W_{32} N_3 - (A_{32} + A_{31}) N_3 \overset{!}{=} 0 \tag{2.20}$$

and thus to

$$\frac{N_3}{N_2} = \frac{\frac{W_p}{f} + W_{32}}{W_{32} + A_{32} + A_{31}} > 1 \quad \forall W_p > \frac{f}{\tau_3}. \tag{2.21}$$

Here, the upper level lifetime corresponds to $\tau_3 = (A_{32} + A_{31})^{-1}$. The laser radiation on the transition $|3\rangle \rightarrow |2\rangle$ now suffers from reabsorption and a population inversion between level $|3\rangle$ and level $|2\rangle$ is only reached for a minimum pump rate. However, owing to $f < 1$ this quasi-three-level laser, will therefore, show a lower threshold than the three-level laser as can be seen from Eq. (2.12).

In a solid-state laser, i.e. a laser based on a medium doped with active ions, the intrinsic levels of the electrons of the dopant ions are split and shifted by the host crystal field. In the case of rare earth ions as, e.g, Nd^{3+}, Er^{3+}, Tm^{3+} or Ho^{3+}, the optically active electron belongs to an inner shell of the ion configuration, and therefore, is shielded from the strong influence of the crystal field. The electronic levels itself are split into different terms resulting from the non-centrosymmetric

Fig. 2.3 Splitting of energy
levels in a solid-state host on
the example of the Ho^{3+} ion
[2]

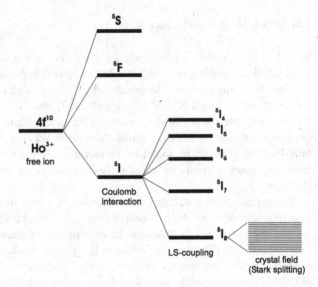

Fig. 2.4 Quasi-three level
scheme in a solid-state laser

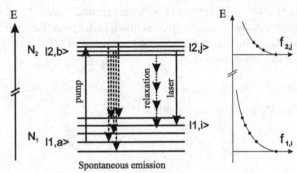

coulomb interaction and the **spin-orbit (LS) coupling** within the ion. The weak-
ened influence of the crystal field then causes the splitting of the different terms as a
result of the **Stark effect** and is shown in Fig. 2.3 on the example of the Ho^{3+} ion.
As the **Stark splitting** energy difference (some $100\ cm^{-1}$) is less than the energy
difference between the main terms (some $1000\ cm^{-1}$), e.g. the $^5I_8$ and $^5I_7$ with
$\sim 5000\ cm^{-1}$ separation, the **Stark levels** form separated groups, called **manifolds**.
In this example, a pump and laser transition would connect different Stark levels in
the $^5I_8$ and $^5I_7$ manifold as sketched in Fig. 2.4.

This manifold structure is common to all rare earth ions in solids, as can be
seen from Fig. 2.5 for rare earth ions in the host $LaCl_3$. As the main energy
shifts arise from the Coulomb and the spin-orbit interaction, which are inherent
processes of the ion itself, whilst the Stark-splitting is comparably weak, this diagram also shows qualitatively the energetic positions of the manifolds in other host
media.

In this Stark manifold picture, we refer N_m to the total population of a manifold m, here as N_1 for the ground-state manifold and N_2 for the upper level man-

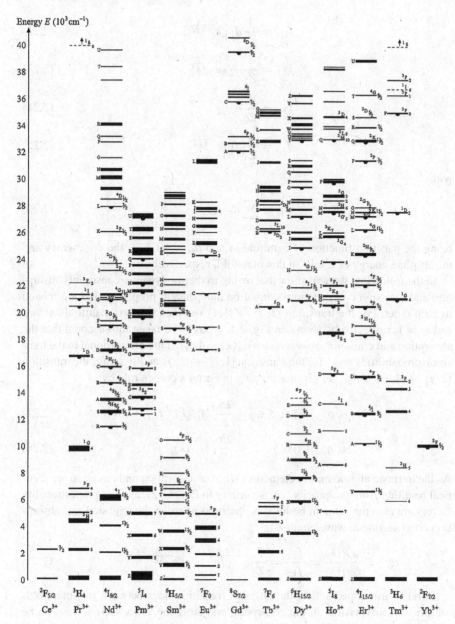

Fig. 2.5 Energy-level diagram of trivalent rare earth ions in LaCl₃ [3, 4]

ifold, whilst the pump radiation couples the levels $|1, a\rangle$ and $|2, b\rangle$ and the laser radiation couples the levels $|2, j\rangle$ and $|1, i\rangle$. As the levels within one manifold are close enough to each other so that a thermal population distribution results within a manifold, the fractional population of the levels itself is given by

$$f_{1,a} = \frac{1}{Z_1} d_{1,a} e^{-\frac{E_{1,a}}{k_B T}}, \tag{2.22}$$

$$f_{2,b} = \frac{1}{Z_2} d_{2,b} e^{-\frac{E_{2,b}}{k_B T}}, \tag{2.23}$$

$$f_{2,j} = \frac{1}{Z_2} d_{2,j} e^{-\frac{E_{2,j}}{k_B T}}, \tag{2.24}$$

$$f_{1,i} = \frac{1}{Z_1} d_{1,i} e^{-\frac{E_{1,i}}{k_B T}}, \tag{2.25}$$

with

$$Z_m = \sum_{i \in m} d_{m,i} e^{-\frac{E_{m,i}}{k_B T}} \tag{2.26}$$

being the partition function of manifold m and $d_{m,i}$ and $E_{m,i}$ the degeneracy and the absolute energy of level i in this manifold, respectively.

In the following, we will show that owing to thermodynamic constraints absorption and emission cross sections cannot be independent properties and are related to each other. For the transition $|2, j\rangle \leftrightarrow |1, i\rangle$ between the two manifolds $m = 2$ and $m = 1$, respectively, shown in Fig. 2.4, it has to be taken into account that the absorption and emission cross sections $\sigma_a(\lambda)$ and $\sigma_e(\lambda)$ are proportional to the transition probabilities $p_{i \to j}$ for the transition $|1, i\rangle \to |2, j\rangle$ and $p_{j \to i}$ for the transition $|2, j\rangle \to |1, i\rangle$, which are given according to Fermi's golden rule by

$$\sigma_{a,ij}(\lambda) \propto p_{i \to j} = \frac{2\pi}{\hbar} |M_{ij}(\lambda)|^2 f_{1,i}, \tag{2.27}$$

$$\sigma_{e,ij}(\lambda) \propto p_{j \to i} = \frac{2\pi}{\hbar} |M_{ji}(\lambda)|^2 f_{2,j}. \tag{2.28}$$

As the intrinsic atomic matrix elements $|M_{ij}|$ for absorption and emission are identical resulting from reciprocity, i.e. according to Eq. (1.12), and the proportionality factors are also the same in both cases, the ratio between the emission and absorption cross section at wavelength λ is

$$\frac{\sigma_{e,ij}(\lambda)}{\sigma_{a,ij}(\lambda)} = \frac{f_{2,j}}{f_{1,i}} = \frac{Z_1}{Z_2} e^{-\frac{E_{2,j} - E_{1,i}}{k_B T}} = e^{-\frac{hc}{k_B T}(\frac{1}{\lambda} - \frac{1}{\lambda_\mu})}. \tag{2.29}$$

In rare-earth doped solids, the different transition lines between two manifolds are often clearly visible in the absorption and emission cross sections, as can be seen in Fig. 2.6. This is a result of the less strong interaction of the inner-shell electrons responsible for the transitions with the crystal field of the host, compared with, for example, transition metal doped solids, for which the outer-shell electrons are responsible for the optical transitions, which are not shielded from the crystal field. The spectroscopic absorption and emission cross sections, i.e. the externally measured cross sections according to Eq. (1.68), can therefore, be written as a sum over the intrinsic atomic cross sections σ_{ij} connecting the different levels,

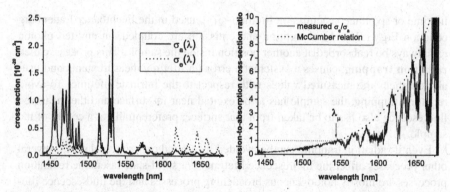

Fig. 2.6 Emission and absorption cross sections of Er^{3+}:YAG and the corresponding ratio compared with the McCumber relation

$$\sigma_a(\lambda) = \sum_{ij} f_{1,i}\sigma_{ij}(\lambda)d_{2,j}, \tag{2.30}$$

$$\sigma_e(\lambda) = \sum_{ij} f_{2,j}\sigma_{ji}(\lambda)d_{1,i}. \tag{2.31}$$

In Eq. (1.68) the N_m refer to the manifold population density, and therefore, the fractional population of each level, as well as the degeneracy of the final level, absorbed into what is called **spectroscopic cross section**. From reciprocity, i.e. Eq. (1.12), it follows for the intrinsic cross sections that $\sigma_{ij}(\lambda) = \sigma_{ji}(\lambda)$.

Using Eq. (2.29), the equivalent relation

$$\frac{\sigma_e(\lambda)}{\sigma_a(\lambda)} = e^{-\frac{hc}{k_B T}(\frac{1}{\lambda} - \frac{1}{\lambda_\mu})} \tag{2.32}$$

can be deduced. This equation is also known as **McCumber relation** and is shown for Er^{3+}:YAG in Fig. 2.6. The exact derivation can be found in [5]. The **chemical potential wavelength** λ_μ can be expressed by

$$\lambda_\mu = \frac{hc}{k_B T}\left(\ln\frac{Z_1}{Z_2}\right)^{-1}. \tag{2.33}$$

Spectroscopically, this quantity is given by the intersection between the spectroscopic absorption and emission cross sections. The McCumber relation allows the determination of the emission cross section $\sigma_e(\lambda)$ from the absorption cross section $\sigma_a(\lambda)$ or can be used as a proof of measured data together with the Füchtbauer-Ladenburg relation, Eq. (1.76). It has to be noted, however, that the spectroscopic measurements often contain large errors, especially when the cross sections are low and the detection noise becomes important. Then a deviation is found between the experimentally measured ratio and the McCumber relation, as can be seen in Fig. 2.6. Nevertheless, for the peaks of the cross sections where detector noise is low, the McCumber relation shows a good agreement with the measurements.

The fluorescence decay time τ_f of an excitation, often also called the (fluorescence) lifetime of a manifold, however, is usually different from the rediative lifetime or spontaneous emission lifetime $\tau_{21,sp}$ used in the Füchtbauer-Ladenburg equation Eq. (1.76). As absorption and emission are coupled, an emitted photon can always be reabsorbed at another position inside the sample. This process, called **radiation trapping**, causes a systematic error on lifetime measurements and usu- aly lengthens the measured values with respect to the intrinsic lifetime. To avoid radiation trapping, the sample has to be excited near its surface and the measured fluorescence also has to be taken from that surface, preferentially at a corner of the sample.

Even if radiation trapping can be avoided by a good experimental setup, several other processes affect the fluorescence lifetime. For solid-state lasers the relaxation processes are mostly homogeneous broadening processes and the fluorescence life- time is, as discussed in Sect. 1.4.2, given by the sum of the transition rates of the radiative and relaxation processes by

$$\frac{1}{\tau_f} = \frac{1}{\tau_{21,sp}} + \frac{1}{\tau_r}, \tag{2.34}$$

where $\tau_{21,sp}$ is the radiative lifetime and τ_r is the **relaxative**, i.e. **non-radiative** life- time. The radiative lifetime $\tau_{21,sp}$ is the average spontaneous lifetime of a manifold $|2\rangle$

$$\frac{1}{\tau_{2,sp}} = \sum_{j \in |2\rangle} \frac{f_{2,j}}{\tau_{2,sp,j}}, \tag{2.35}$$

so the intrinsic lifetime of level j inside the manifold $|2\rangle$ can be expressed by

$$\frac{1}{\tau_{2,sp,j}} = 8\pi n^2 c \sum_{i \in |1\rangle} \int \frac{\sigma_{ji}(\lambda) d_i}{\lambda^4} d\lambda. \tag{2.36}$$

At small ion concentrations the non-radiative contribution is mainly given by **multi-phonon relaxation**, in which the energy gap ΔE between the two manifolds is bridged by the emission of $n_P \approx \frac{\Delta E}{E_P}$ phonons to the crystal lattice. E_P is the dominating phonon energy. When the crystal lattice itself has a phonon occupation number n^*, this results in a **multi-phonon transition rate** W_{MP} given by [6]

$$W_{MP} = \frac{1}{\tau_r} = W_0 (n^* + 1)^{n_P}. \tag{2.37}$$

Therein W_0 is the spontaneous multi-phonon transition rate at $T = 0$ K as a result of the zero-point fluctuations of the phonon field. Inserting the Boson occupation average $\langle n^* \rangle = \frac{1}{e^{\frac{E_P}{k_B T}} - 1}$ results in

$$W_{MP} = \frac{1}{\tau_r} = \frac{W_0}{(1 - e^{-\frac{E_P}{k_B T}})^{n_P}}. \tag{2.38}$$

The spontaneous multi-phonon transition rate W_0 is often expressed in the form

$$W_0 = B e^{-a \Delta E}, \tag{2.39}$$

wherein B and a are material-dependent parameters. They have to be determined empirically. Resulting from the exponential dependence on the energy gap only the multi-phonon relaxation from a level or manifold to the next lower level or manifold has to be considered.

For the processes discussed so far, the measured fluorescence lifetime is independent of the excitation density and the total active ion density of the sample under non-lasing conditions and when radiation trapping can be neglegted. Density dependent fluorescence lifetimes can occur and are connected with energy transfer processes that will be discussed in Sect. 5.2.2.

Whilst on the one hand the low separation between the lower laser level and the absolute ground state $|1, 0\rangle$ in a quasi-three-level laser negatively affects the laser performance as a result of the increased reabsorption, it allows highly efficient operation in terms of relative photon energy as the conversion efficiency from pump to laser signal photon energies of one closed-cycle transition, the **quantum efficiency** η_Q, will be given by the ratio of the photon energies

$$\eta_Q = \frac{v_s}{v_p} = \frac{\lambda_p}{\lambda_s}. \tag{2.40}$$

In this equation, the indices s and p denote the laser signal and pump radiation.

2.1.3 The Feedback Condition

Up to now we have investigated the possibility of the creation of a population inversion, which is a necessary condition in order to realize an effective gain of $G > 1$ in a laser medium. It is a well known fact that a gain element can be turned into a self-oscillating system by adding feedback to this element. Thus, to turn the optical gain medium into a laser, an **optical resonator** needs to be added that reflects the amplified radiation back and forth into the gain medium as shown in Fig. 2.7. This resonator is often also called **laser cavity**.

Therefore, two mirrors are added at opposite ends of the medium, from which one has a high reflection at the laser wavelength (HR mirror) whilst the other one has a partial reflection at the laser wavelength ($R_{OC} < 1$). This partially reflective mirror is called **output coupler** (OC mirror), as it allows the internal radiation to leave the resonator. From Fig. 2.7 the necessary population inversion of a laser oscillator can be deduced. In this description, all internal passive losses of the resonator are summed into the loss parameter Λ and are taken to occur at the HR mirror. From self-consistency of the internal laser intensity after one round-trip starting at the OC mirror in the left direction, we can deduce that

$$G \times R_{HR} \times (1 - \Lambda) \times G \times R_{OC} = G^2 R_{HR}(1 - \Lambda) R_{OC} = 1, \tag{2.41}$$

resulting in a necessary single-pass gain G of

$$G = \frac{1}{\sqrt{R_{HR}(1 - \Lambda) R_{OC}}}. \tag{2.42}$$

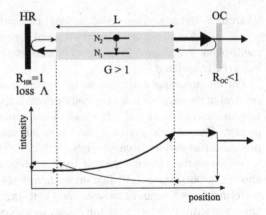

Fig. 2.7 Schematic representation of the feedback in a laser and the resulting self-consistent intensity distribution

In the following, we will relate this gain equation to the manifold population densities of a laser medium in which the gain occurs from an effective population inversion between the manifolds $|2\rangle$ and $|1\rangle$.

As the intensity varies locally caused by the gain in the laser medium, we have to take into account axially varying population densities. Nevertheless, in order to allow a simple description, the axial average along the laser axis

$$\langle \cdot \rangle = \frac{1}{L} \int_0^L \cdot \, dz \tag{2.43}$$

is introduced. Under the assumption that no other manifolds are involved, i.e. $N_1 + N_2 = N$, Eq. (2.2) can be used and the corresponding population densities are given by

$$\langle N_2 \rangle = \frac{\sigma_a(\lambda_s)}{\sigma_a(\lambda_s) + \sigma_e(\lambda_s)} \langle N \rangle + \frac{1}{L} \frac{\ln G}{\sigma_a(\lambda_s) + \sigma_e(\lambda_s)}, \tag{2.44}$$

$$\langle N_1 \rangle = \frac{\sigma_e(\lambda_s)}{\sigma_a(\lambda_s) + \sigma_e(\lambda_s)} \langle N \rangle - \frac{1}{L} \frac{\ln G}{\sigma_a(\lambda_s) + \sigma_e(\lambda_s)}, \tag{2.45}$$

where λ_s is the emission wavelength of the laser. This is the second condition that has to be fulfilled in order to create a laser, stating that an effective population inversion according to Eq. (2.3) is a necessary, but not sufficient condition for laser operation. Furthermore, the internal losses and the outcoupling in the laser also need to be compensated for. The population density ratio therefore results in

$$\frac{\langle N_2 \rangle}{\langle N_1 \rangle} = \frac{\sigma_a(\lambda_s) \langle N \rangle L + \ln G}{\sigma_e(\lambda_s) \langle N \rangle L - \ln G}. \tag{2.46}$$

In this notation a true four-level laser is described by $\sigma_a(\lambda_s) = 0$ whilst for a true three-level laser one finds $\sigma_a(\lambda_s) = \sigma_e(\lambda_s)$, showing again that at least half of the population needs to be in level $|2\rangle$. Then the population densities N_i refer to the single levels.

2.2 Spectroscopic Laser Rate Equations

In the following discussion, we will set up the laser rate equations including the photon field of the resonator to investigate the general behaviour of a laser oscillator in **continuous wave (cw)** operation in Sect. 2.2.1. This corresponds to the stationary regime of these rate equations. Whenever this stationary solution is perturbed by external influences or at the start of the laser, relaxation oscillations occur that will be described in Sect. 2.2.2. We will use the spectroscopic laser rate equations as in Sect. 2.1.3 to describe this condition. The difference between the three- and four-level laser is thus, again, given by $\sigma_a(\lambda_s) = \sigma_e(\lambda_s)$ and $\sigma_a(\lambda_s) = 0$, respectively, with the quasi-three-level laser being in between of these two limits. In Sect. 2.3 a potential model of a laser is presented that describes the laser field intensity and phase and can thus explain the coherence properties of the laser field.

2.2.1 Population and Stationary Operation

In this description, we refer to the manifold scheme in Fig. 2.4. In order to derive the rate equation for the upper laser manifold population density N_2 the relations given in the Eqs. (1.70), (1.71) are used. This results in

$$\left(\frac{\partial N_2}{\partial t}\right)_{em} = -\int \frac{1}{h\nu}\frac{\partial \tilde{P}}{\partial V}d\nu = -N_2 \int \frac{\sigma_e(\nu)\tilde{I}(\nu)}{h\nu}d\nu = -\frac{N_2}{hc}\int \lambda\sigma_e(\lambda)\tilde{I}(\lambda)d\lambda \tag{2.47}$$

for the change of N_2 per unit time caused by stimulated emission. For simplicity reasons we assume a single laser frequency $\tilde{I}(\lambda) = I_s\delta(\lambda - \lambda_s)$. In this case, the rate equation for stimulated emission is given by

$$\left(\frac{\partial N_2}{\partial t}\right)_{em} = -\frac{\lambda_s}{hc}\sigma_e(\lambda_s)I_s N_2 = -W_{21}N_2. \tag{2.48}$$

Therefore, it can be concluded that the transition rate W_{21} is proportional to the incident intensity on the transition wavelength and is given by

$$W_{21} = \frac{\lambda_s}{hc}\sigma_e(\lambda_s)I_s. \tag{2.49}$$

The corresponding transition rate of the reverse process, the absorption rate W_{12}, is determined by analogy with the emission rate, resulting in

$$W_{12} = \frac{\lambda_s}{hc}\sigma_a(\lambda_s)I_s. \tag{2.50}$$

In the following discussion, a longitudinally pumped laser medium is analyzed as shown in Fig. 2.8. The full rate equations for the population densities of this longitudinally pumped laser medium under the assumption $N_2 + N_1 = N$, including pump intensity I_p, laser intensity I_s and spontaneous decay τ, are thus given by

Fig. 2.8 Setup of a
longitudinally pumped laser
medium with integrated
mirrors

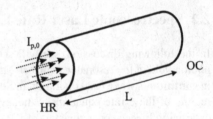

Fig. 2.9 Representation of
the intensity I_s as a constant
photon density Φ in the
volume $V = Al$

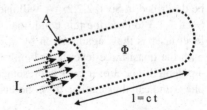

$$\frac{\partial N_2}{\partial t} = \frac{\lambda_p}{hc}\alpha(z)I_{p,0}e^{-\int_0^z \alpha(z')dz'} + \frac{\lambda_s}{hc}I_s\left[\sigma_a(\lambda_s)N_1 - \sigma_e(\lambda_s)N_2\right] - \frac{N_2}{\tau}$$

$$\frac{\partial N_1}{\partial t} = -\frac{\partial N_2}{\partial t},$$

(2.51)

where, $\alpha = \sigma_a(\lambda_p)N_1 - \sigma_e(\lambda_p)N_2$ is the pump absorption coefficient and $I_{p,0}$ the incident pump intensity that decreases along the propagation axis z as

$$I_p = I_{p,0}e^{-\int_0^z \alpha(z')dz'}.$$

(2.52)

Introducing the pump absorption efficiency

$$\eta_{abs} = 1 - e^{-\int_0^L \alpha(z')dz'},$$

(2.53)

the rate equations averaged over the laser axis can be written as

$$\frac{\partial \langle N_2 \rangle}{\partial t} = \frac{\lambda_p}{hc}I_p\frac{\eta_{abs}}{L} + \frac{\lambda_s}{hc}\left[\sigma_a(\lambda_s)\langle I_s N_1 \rangle - \sigma_e(\lambda_s)\langle I_s N_2 \rangle\right] - \frac{\langle N_2 \rangle}{\tau}$$

$$\frac{\partial \langle N_1 \rangle}{\partial t} = -\frac{\partial \langle N_2 \rangle}{\partial t}.$$

(2.54)

By analogy, we will now set up a rate equation for the photon density in the cavity. The photon density Φ is directly connected to the field intensity I_s. To deduce this relation we refer to Fig. 2.9. A laser intensity I_s that strikes the surface A perpendicularly will propagate a length $l = ct$ during the time t. This corresponds to a total energy E_s included in the volume $V = Al$ given by

$$E_s = I_s At = I_s A \frac{l}{c} = \frac{I_s}{c}V \overset{!}{=} h\nu\Phi V.$$

(2.55)

Thus, we can deduce the photon density as

$$\Phi = \frac{\lambda_s}{hc^2}I_s.$$

(2.56)

The photon density in the resonator will change as a result of the stimulated emission and absorption as well as resulting from losses and outcoupling (useful emission from the cavity), which can be described by a cavity photon lifetime τ_c. This results in

$$\frac{\partial \Phi}{\partial t} = W_{21}N_2 - W_{12}N_1 - \frac{\Phi}{\tau_c}. \tag{2.57}$$

Including Eq. (2.56) yields

$$\frac{\partial \Phi}{\partial t} = c[\sigma_e(\lambda_s)N_2 - \sigma_a(\lambda_s)N_1]\Phi - \frac{\Phi}{\tau_c}, \tag{2.58}$$

or as an axial average

$$\frac{\partial \langle \Phi \rangle}{\partial t} = c[\sigma_e(\lambda_s)\langle \Phi N_2 \rangle - \sigma_a(\lambda_s)\langle \Phi N_1 \rangle] - \frac{\langle \Phi \rangle}{\tau_c}. \tag{2.59}$$

As we assume a constant total population $N = N_2 + N_1$, i.e. no population in other manifolds than $|2\rangle$ and $|1\rangle$, we can transform the rate equations (2.54), (2.59) by introducing a new variable, the inversion $\Delta N = N_2 - N_1$. This results in

$$\frac{\partial \langle \Delta N \rangle}{\partial t} = 2\frac{\lambda_p}{hc}I_p\frac{\eta_{abs}}{L}$$
$$+ c([\sigma_a(\lambda_s) - \sigma_e(\lambda_s)]\langle \Phi N \rangle - [\sigma_a(\lambda_s) + \sigma_e(\lambda_s)]\langle \Phi \Delta N \rangle)$$
$$- \frac{\langle N \rangle + \langle \Delta N \rangle}{\tau} \tag{2.60}$$

$$\frac{\partial \langle \Phi \rangle}{\partial t} = \frac{c}{2}([\sigma_a(\lambda_s) + \sigma_e(\lambda_s)]\langle \Phi \Delta N \rangle - [(\sigma_a(\lambda_s) - \sigma_e(\lambda_s)]\langle \Phi N \rangle)$$
$$- \frac{\langle \Phi \rangle}{\tau_c}. \tag{2.61}$$

The laser dynamics are thus described by two time-dependent variables only, the average inversion density $\langle \Delta N \rangle(t)$ and the average photon density $\langle \Phi \rangle(t)$. The term $\langle N \rangle$ is explicitly not simplified to N because in this form the rate equations also allow the description of laser media with axially varying dopant concentrations.

In order to derive some important properties of a laser, we may rewrite Eq. (2.61) in the form

$$\left\langle \frac{1}{\Phi}\frac{\partial \Phi}{\partial t}\right\rangle = \left\langle \frac{c}{2}\left([\sigma_a(\lambda_s) + \sigma_e(\lambda_s)]\Delta N - [(\sigma_a(\lambda_s) - \sigma_e(\lambda_s)]N - \frac{1}{\tau_c}\right)\right\rangle. \tag{2.62}$$

As laser operation can only occur for $\frac{\partial \langle \Phi \rangle}{\partial t} \geq 0$, and owing to $\Phi > 0$, we can deduce the necessary condition

$$[\sigma_a(\lambda_s) + \sigma_e(\lambda_s)]\langle \Delta N \rangle - [\sigma_a(\lambda_s) - \sigma_e(\lambda_s)]\langle N \rangle \geq \frac{2}{c\tau_c}, \tag{2.63}$$

stating that the number of generated photons per unit time must at least compensate for the number of photons per unit time that are lost in the cavity and emitted through the outcoupling mirror. In the case of the equality in Eq. (2.63) this relation is consistent with the round-trip condition, Eqs. (2.44), (2.45), and allows the deduction of the cavity photon lifetime τ_c, resulting in

$$\tau_c = \frac{L}{c\ln G} = -\frac{2L}{c\ln[R_{OC}(1 - \Lambda)R_{HR}]}. \tag{2.64}$$

Fig. 2.10 Population inversion density and laser output intensity, as a function of the pump intensity below and above threshold

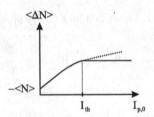
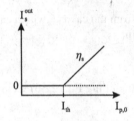

Stationary Operation

In stationary operation, i.e. for $\frac{\partial \langle \Delta N \rangle}{\partial t} = 0$ and $\frac{\partial \langle \Phi \rangle}{\partial t} = 0$, the Eqs. (2.60), (2.61) can be solved for $\langle \Delta N \rangle$ and $\langle \Phi \rangle$, resulting in two sets of solutions. The first one is given by

$$\langle \Delta N \rangle = 2 \frac{\lambda_p \tau}{hc} I_{p,0} \frac{\eta_{abs}}{L} - \langle N \rangle \tag{2.65}$$

$$\langle \Phi \rangle = 0 \tag{2.66}$$

and describes the pumping of the laser below threshold. In this regime, the population inversion increases with pump intensity until the threshold pump intensity I_{th} is reached, as shown in Fig. 2.10. It has to be noted that Eq. (2.65) is a transcendental equation caused by the dependence of $\eta_{abs}(\langle \Delta N \rangle)$, explicitly written as

$$\langle \Delta N \rangle = 2 \frac{\lambda_p \tau}{hc} \frac{I_{p,0}}{L} \left(1 - e^{-[\sigma_a(\lambda_p) - \sigma_e(\lambda_p)]\langle N \rangle L + [\sigma_a(\lambda_p) + \sigma_e(\lambda_p)]\langle \Delta N \rangle L} \right) - \langle N \rangle. \tag{2.67}$$

Therefore, only numerical solutions exist for the behaviour below threshold.

For $I_p > I_{th}$ the first solution becomes unstable and the second solution describes the laser process,

$$\langle \Delta N \rangle = \langle \Delta N \rangle_{th} = \frac{2 \ln G}{[\sigma_a(\lambda_s) + \sigma_e(\lambda_s)]L} + \frac{\sigma_a(\lambda_s) - \sigma_e(\lambda_s)}{\sigma_a(\lambda_s) + \sigma_e(\lambda_s)} \langle N \rangle \tag{2.68}$$

$$\langle \Phi \rangle = \frac{\lambda_p}{hc^2} \frac{\eta_{abs}}{\ln G} (I_p - I_{th}). \tag{2.69}$$

The threshold pump intensity results in

$$I_{th} = \frac{I_{sat}^p}{\eta_{abs}} \left(\ln G + \sigma_a(\lambda_s) \langle N \rangle L \right), \tag{2.70}$$

with the pump saturation intensity

$$I_{sat}^p = \frac{hc}{\lambda_p [\sigma_a(\lambda_s) + \sigma_e(\lambda_s)]\tau}. \tag{2.71}$$

The transition between both regimes occurs at that point where both solutions result in the same average inversion density $\langle \Delta N \rangle$, which itself is given by the threshold pump intensity I_{th}. As can be seen from Eq. (2.68) the inversion density $\langle \Delta N \rangle$ is clamped at its threshold level $\langle \Delta N \rangle_{th}$ and any further increase in input power is transfered into the output. The output intensity I_s^{out} of the laser is obtained from the photon outcoupling lifetime using Eq. (2.64) considering the useful outcoupling only,

$$\tau_{out} = -\frac{2L}{c \ln R_{OC}}, \tag{2.72}$$

and using Eq. (2.55) as

$$I_s^{out} = \frac{P_s^{out}}{A} = \frac{E_s}{\tau_{out}A} = \frac{I_s AL}{c\tau_{out}A} = -\frac{\ln R_{OC}}{2} I_s. \tag{2.73}$$

From Eq. (2.56) the important output-to-input intensity relation of the laser can be deduced,

$$I_s^{out} = \frac{\lambda_p}{\lambda_s} \frac{-\ln R_{OC}}{2 \ln G} \eta_{abs}(I_p - I_{th}), \tag{2.74}$$

or in terms of power

$$P_s^{out} = \frac{\lambda_p}{\lambda_s} \frac{-\ln R_{OC}}{2 \ln G} \eta_{abs}(P_p - P_{th}), \tag{2.75}$$

The **slope efficiency** describes the differential slope of the output versus input power of a laser and is thus given by

$$\eta_s = \frac{\lambda_p}{\lambda_s} \frac{-\ln R_{OC}}{2 \ln G} \eta_{abs} = \frac{\lambda_p}{\lambda_s} \frac{-\ln (1 - T_{OC})}{2 \ln G} \eta_{abs}, \tag{2.76}$$

which includes the quantum efficiency η_Q of Eq. 2.40, the output coupler transmission $T_{OC} = 1 - R_{OC}$ and the pump absorption efficiency η_{abs}. The term $2 \ln G$ corresponds to the total losses of the laser cavity, see Eq. (2.42), and $-\ln (1 - T_{OC})$ accounts for the useful outcoupling losses. For small outcoupling $T_{OC} \ll 1$ one often finds the approximation $-\ln (1 - T_{OC}) \approx T_{OC}$.

In order to achieve the maximum efficiency for a given laser the amount of cavity emission from the outcoupling mirror, or the output coupler reflectivity R_{OC}, needs to be optimized with respect to the relevant cavity conditions. A high value of outcoupling will result in a high slope efficiency, but also in a high threshold, whilst a low value of outcoupling, i.e. a high reflectivity of the OC mirror, will strongly enhance the intra-cavity circulating intensity, and therefore, increase the effect of the internal losses Λ. This optimization usually needs to be undertaken for a given laser medium as the absorption efficiency, which is a function of the inversion and thus of the gain G,

$$\eta_{abs} = 1 - e^{2 \frac{\sigma_a(\lambda_p)+\sigma_e(\lambda_p)}{\sigma_a(\lambda_s)+\sigma_e(\lambda_s)} \ln G} e^{(\sigma_e(\lambda_p)-\sigma_a(\lambda_p)+\frac{\sigma_a(\lambda_p)+\sigma_e(\lambda_p)}{\sigma_a(\lambda_s)+\sigma_e(\lambda_s)}[\sigma_a(\lambda_s)-\sigma_e(\lambda_s)])\langle N\rangle L}, \tag{2.77}$$

includes the spectroscopic properties of the laser medium. However, even if we do not refer to a special laser medium, we can deduce the fundamental ideas of this optimization. Therefore, we neglect the dependence of η_{abs} and find from Eq. (2.76) that $\frac{-\ln(1-T_{OC})}{2\ln G}$ should be optimized in order to get a high slope efficiency. In order to get a low pump threshold, we must reduce $\ln G$ and, in a quasi-three-level laser medium, the re-absorption $\sigma_a(\lambda_s)\langle N\rangle L$, i.e. the length-concentration product

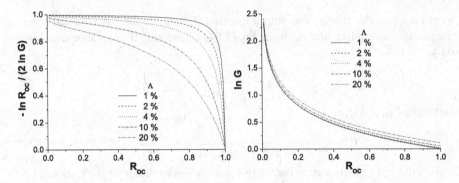

Fig. 2.11 $\frac{-\ln R_{OC}}{2\ln G}$ and $\ln G$ as a function of the output coupler reflectivity R_{OC} and the internal cavity losses Λ

Fig. 2.12 Comparison of the different laser schemes

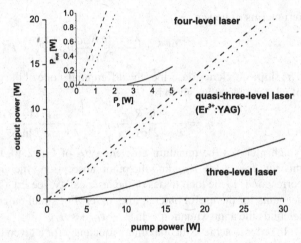

of the laser medium, as follows from Eq. (2.70). However, the main influence on the threshold is given by the term $\ln G$. As can be seen in Fig. 2.11, the intra-cavity losses mainly influence the slope efficiency whilst having a nearly negligible effect on the threshold.

As a comparison, the input-to-output power curves of the three different laser schemes are shown in Fig. 2.12. These are results of numerical calculations, which are a bit more complex that the description discussed above [7]. Therein, the spectroscopic properties of a real existing Er^{3+}:YAG laser emitting at $\lambda_s = 1645$ nm are used for the model of the quasi-three-level laser. These data are then modified to describe a hypothetical four-level and three-level laser with the same properties, i.e. for the four-level laser the absorption cross section $\sigma_a(\lambda_s)$ is artificially set to zero, whilst for the three-level laser the absorption cross section $\sigma_a(\lambda_s)$ is artificially set to equal the emission cross section $\sigma_e(\lambda_s)$. As already discussed previously, the four-level laser shows the lowest threshold and the highest slope efficiency whereas the three-level laser has the highest threshold and lowest slope efficiency. The non-linear start of the three-level and quasi-three-level laser comes from the effect that the additional reabsorption loss in the cavity is saturated for high intracavity laser intensities. Thus, quasi-three-level lasers show an increasing slope efficiency with pump power that converges towards the four-level laser slope efficiency for very high laser powers. The real quasi-three-level laser is found in between the two limits, being as closer to a four-level laser for lower re-absorption cross sections $\sigma_a(\lambda_s) \ll \sigma_e(\lambda_s)$.

2.2.2 Relaxation Oscillations

As a result of the mixed products $\langle \Phi \Delta N \rangle$ in the rate equations (2.60), (2.61) no analytic solution can be given for the full temporal behaviour of the laser. It is, however, possible to investigate the temporal behaviour of these equations for infinitesimal deviations from the steady-state solution given in Eqs. (2.68), (2.69). These deviations occur, for example, as a result of external perturbations like vibrations of the mirrors or the laser setup itself. To calculate these effects we have to **linearize** the rate equations and write the inversion density $\langle \Delta N \rangle(t)$ and the photon density $\langle \Phi \rangle(t)$ in the form

$$\langle \Delta N \rangle(t) = \langle \Delta N_{th} \rangle + \langle \Sigma \rangle(t) \tag{2.78}$$

$$\langle \Phi \rangle = \langle \Phi_{cw} \rangle + \langle \Pi \rangle(t), \tag{2.79}$$

with $\langle \Sigma \rangle(t) \ll \langle \Delta N_{th} \rangle$ and $\langle \Pi \rangle(t) \ll \langle \Phi_{cw} \rangle$, where, $\langle \Phi_{cw} \rangle$ denotes the steady-state solution of Eq. (2.69). As a simplification, we additionally assume that the inversion density shows no strong axial dependence, so that any averaged product can be written as the product of the averages, e.g. $\langle \Phi \Delta N \rangle \approx \langle \Phi \rangle \langle \Delta N \rangle$. By inserting Eqs. (2.78), (2.79) into the rate equations (2.60), (2.61), neglecting doubly infinitesimal terms such as $\langle \Pi \rangle \langle \Sigma \rangle$, the **linearized rate equations**

$$\frac{\partial \langle \Sigma \rangle}{\partial t} = c\big[\sigma_a(\lambda_s) - \sigma_e(\lambda_s)\big]\langle \Pi \rangle \langle N \rangle$$

$$- c\big[\sigma_a(\lambda_s) + \sigma_e(\lambda_s)\big]\langle \Pi \rangle \langle \Delta N_{th} \rangle$$

$$- c\big[\sigma_a(\lambda_s) + \sigma_e(\lambda_s)\big]\langle \Phi_{cw} \rangle \langle \Sigma \rangle - \frac{\langle \Sigma \rangle}{\tau} \tag{2.80}$$

$$\frac{\partial \langle \Pi \rangle}{\partial t}\bigg|_{\langle \Pi \rangle = 0} = \frac{c}{2}\big[\sigma_a(\lambda_s) + \sigma_e(\lambda_s)\big]\langle \Phi_{cw} \rangle \langle \Sigma \rangle, \tag{2.81}$$

result, where, the equation for the photon density is linearized around $\langle \Pi \rangle = 0$. By taking the derivative of Eq. (2.81) with respect to time and inserting Eq. (2.80), the product terms can be eliminated, which yields

$$\frac{\partial^2 \langle \Pi \rangle}{\partial t^2} + 2\xi \frac{\partial \langle \Pi \rangle}{\partial t} + \Omega_0^2 \langle \Pi \rangle = 0. \tag{2.82}$$

Fig. 2.13 Calculated normalized amplitude $\frac{\Pi(t)}{\hat{\Pi}}$ of the relaxation oscillations of an Er^{3+}:YAG laser. The inset shows an enlarged view on the 100 µs scale

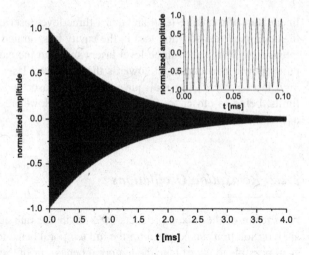

This is the equation of a damped harmonic oscillator, stating that the photon density will oscillate around its steady state value with a frequency $\Omega = \sqrt{\Omega_0^2 - \xi^2}$ and an amplitude that is damped with the time constant ξ as

$$\langle \Pi \rangle(t) = \hat{\Pi} e^{-\xi t} \cos \Omega t. \tag{2.83}$$

The corresponding time constants are found to be

$$\xi = \frac{1}{2\tau}\left[\left(1 + \sigma_a(\lambda_s)\langle N \rangle c \tau_c\right)\left(\frac{I_p}{I_{th}} - 1\right) + 1\right] \tag{2.84}$$

$$\Omega_0 = \sqrt{\frac{1}{\tau}\left(\frac{1}{\tau_c} + \sigma_a(\lambda_s)\langle N \rangle c\right)\left(\frac{I_p}{I_{th}} - 1\right)}, \tag{2.85}$$

which simplify for a four-level laser ($\sigma_a(\lambda_s) = 0$) to

$$\xi = \frac{1}{2\tau}\frac{I_p}{I_{th}} \tag{2.86}$$

$$\Omega_0 = \sqrt{\frac{1}{\tau\tau_c}\left(\frac{I_p}{I_{th}} - 1\right)}. \tag{2.87}$$

As an example we investigate these values for an Er^{3+}:YAG laser with the parameters $\tau = 7.64$ ms, $R_{OC} = 80$ %, $\Lambda = 2$ %, $L = 0.06$ m, $\sigma_a(1645 \text{ nm}) = 6.67 \times 10^{-22}$ cm^2, $\langle N \rangle = 6.9 \times 10^{25}$ m^{-3} and $I_p = 5I_{th}$, i.e. $G = 1.129$ and $\tau_c = 1.65$ ns. This results in

$$\xi = 923 \text{ s}^{-1} \tag{2.88}$$

$$\Omega_0 = 2\pi \times 163 \text{ kHz}, \tag{2.89}$$

and corresponds to an oscillation frequency of ~ 163 kHz and a damping time of $\xi^{-1} = 1.08$ ms. The relaxation oscillation is shown in Fig. 2.13.

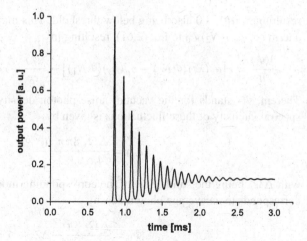

Fig. 2.14 Calculated "spiking" of a solid-state laser under continuous pumping starting at $t = 0$

Spiking

In a special form the relaxation oscillations occur always at the beginning of laser oscillation, i.e. when a laser is turned on. This behaviour, which is given by the transition between the two solutions below and above threshold, i.e. the transition from $\langle \Phi \rangle = 0$ to $\langle \Phi \rangle = \langle \Phi_{cw} \rangle$, differs from the standard relaxation oscillations, described above, because the deviation from steady-state is not small compared with the steady-state value. Therefore, in some texts this behaviour is especially refered to as **spiking**. Spiking is always superimposed on the standard relaxation oscillations whenever a strong external perturbation affected the laser. However, as a result of the strong deviation from the steady-state the spiking cannot be calculated by the linearized rate equations and a numerical solution needs to be used. The result of such a calculation is shown in Fig. 2.14.

Resulting from strong pumping of the laser the inversion rapidly increases below threshold whilst the cavity has no internal photon field $\langle \Phi \rangle \approx 0$. As the photon field needs at least several round-trips in oder to build up from the fluorescence noise, it reacts to a change in population density and thus to a change in gain only with a certain time delay. Thus, an inversion density $\langle \Delta N \rangle > \langle \Delta N_{th} \rangle$ can build up for a short time after the pump is switched on, causing an exponential growth of the internal cavity field, which strongly depletes the inversion. Thus, the laser may even stop oscillating until a new cycle begins creating a second spike. As in each cycle an increasing residual inversion is left after the spike, the spikes decrease in intensity and the output power converges to the steady-state value after several cycles of spikes.

There is still a question to be answered regarding Eqs. (2.60), (2.61): how can the system make the transition from the first solution below threshold to the second solution above threshold when the first solution provides $\langle \Phi \rangle = 0$ for all pump powers below threshold. The answer is the existence of the spontaneous emission, resulting in $\langle \Phi \rangle > 0$ also being below threshold. This may be expressed by adding a term $c\sigma_e(\lambda_s)\langle N_2 \rangle \Phi_0$ to Eq. (2.61), resulting in

$$\frac{\partial \langle \Phi \rangle}{\partial t} = c\left[\sigma_e(\lambda_s)\langle \Phi N_2 \rangle - \sigma_a(\lambda_s)\langle \Phi N_1 \rangle\right] - \frac{\langle \Phi \rangle}{\tau_c} + c\sigma_e(\lambda_s)\langle N_2 \rangle \Phi_0. \qquad (2.90)$$

Therein, Φ_0 stands for the vacuum noise photon density in the laser mode. The spectral intensity of these fluctuations is given by

$$\tilde{I}_{s,0} = \frac{\Delta \Omega_s}{4\pi} \frac{8\pi n^2 h c^2}{\lambda_s^5}, \qquad (2.91)$$

with $\Delta \Omega_s$ being the solid angle of the corresponding mode. Using Eq. (2.56), this corresponds to an average photon density of

$$\Phi_0 = \frac{\Delta \Omega_s}{4\pi} \frac{8\pi n^2}{\lambda_s^4} \Delta \lambda_s \qquad (2.92)$$

in the wavelength bandwidth $\Delta \lambda_s$. Then there is always a small but non-negligible number of photons in the cavity that cause the laser to switch to the second solution, as soon as the pump threshold is passed.

2.3 Potential Model of the Laser

Up to now we have only investigated the fundamental laser properties in terms of the interaction between the population densities of the different levels and the photons of the intra-cavity laser beam. This means that we described the laser medium by its material properties and the laser radiation by its intensity. The laser radiation can also be described by an electromagnetic wave, which itself is given by an intensity and a phase. However, the above rate equation description only accounts for the intensity. Therefore, we will investigate the phase properties of laser light in the following discussion using a slightly different model, which describes the laser field as being analogous to to a particle in a potential well [1].

The laser field in the laser resonator shown in Fig. 2.7 can be described by a standing wave between the HR and OC mirror. We thus assume a linearly polarized wave with a complex electric field amplitude given by

$$E(t, z) = E(t)e^{-i\omega t} \sin kz, \qquad (2.93)$$

with

$$k = \frac{\pi}{L}s, \quad s \in \mathbb{N}. \qquad (2.94)$$

The temporal evolution of the complex amplitude $E(t)$ can be devided into the time-dependent amplitude $\hat{E}(t)$ and the phase $\varphi(t)$ as

$$E(t) = \hat{E}(t)e^{i\varphi(t)}. \qquad (2.95)$$

The rate equation for the electric field amplitude equivalent to the photon density rate equation (2.58) is

$$\frac{\partial E(t)}{\partial t} = \frac{c}{2}\left[\sigma_e(\lambda_s)N_2 - \sigma_a(\lambda_s)N_1\right]E(t) - \frac{E(t)}{2\tau_c}, \qquad (2.96)$$

which can be easily prooved using the relation $I_s(t) = \sqrt{\frac{\epsilon_0}{\mu_0}}|E(t)|^2$ and Eq. (2.56), taking into account that

$$\frac{\partial I_s(t)}{\partial t} = \sqrt{\frac{\epsilon_0}{\mu_0}}\left(E^*(t)\frac{\partial E(t)}{\partial t} + E(t)\frac{\partial E^*(t)}{\partial t}\right). \qquad (2.97)$$

We introduce a term $\epsilon(t)$, which accounts for the spontaneous emission of the laser medium into this equation as in Eq. (2.90). Thus, it adds a time-dependent statistical fluctuation to the electric field, resulting in

$$\frac{\partial E(t)}{\partial t} = \frac{c}{2}\left[\sigma_e(\lambda_s)N_2 - \sigma_a(\lambda_s)N_1\right]E(t) - \frac{E(t)}{2\tau_c} + \epsilon(t). \qquad (2.98)$$

In order to obtain the final equation for the electric field in this potential model, the dependence of the inversion term $g = c[\sigma_e(\lambda_s)N_2 - \sigma_a(\lambda_s)N_1]$ from the field amplitude needs to be included. We can simplify this term in the following way: The population densites N_2 and N_1 will be determined on one side by the optical pumping (marked by ↑), which would result in an inversion term

$$g_\uparrow = c\left[\sigma_e(\lambda_s)N_{2,\uparrow} - \sigma_a(\lambda_s)N_{1,\uparrow}\right] \qquad (2.99)$$

in the absence of a laser field. On the other side, the inversion will be reduced (marked by ↓), i.e. saturated, by the laser field and this effect may be approximated as being proportional to the laser field intensity, as long as this intensity is not too high. Thus, the inversion term g can be written under these assumptions as

$$g = g_\uparrow - g_\downarrow = g_\uparrow - \zeta|E(t)|^2, \qquad (2.100)$$

with an appropriate proportionality constant ζ. This leads to the fundamental equation of the potential model,

$$\frac{\partial E(t)}{\partial t} = \frac{1}{2}(g_\uparrow - A_{21})E(t) - \zeta|E(t)|^2 E(t) + \epsilon(t). \qquad (2.101)$$

Using a mathematical "trick", we can illustrate the meaning of Eq. (2.101) by a mechanical example: By adding a term $m\frac{\partial^2 E(t)}{\partial t^2}$, to the left hand side of Eq. (2.101) we get

$$m\frac{\partial^2 E(t)}{\partial t^2} + \frac{\partial E(t)}{\partial t} = \frac{1}{2}(g_\uparrow - A_{21})E(t) - \zeta|E(t)|^2 E(t) + \epsilon(t), \qquad (2.102)$$

which is identical to the equation of motion of a particle of mass m

$$m\frac{\partial^2 x}{\partial t^2} + \frac{\partial x}{\partial t} = F(x) + \epsilon(t), \qquad (2.103)$$

wherein $F(x) = -\frac{dV}{dx}$ describes the force created by the potential $V(x)$ and $\epsilon(t)$ describes external forces acting on that particle. The "mass" m of this particle has

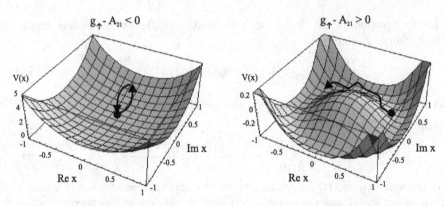

Fig. 2.15 Potential $V(x)$ for the two different cases below and above threshold and the corresponding evolution of the laser field

of-course to be very small in order not to affect the main structure of the equation, i.e. $m\frac{\partial^2 E(t)}{\partial t^2} \ll \frac{\partial E(t)}{\partial t}$. Identifying x with $E(t)$, the corresponding potential $V(x)$ can be expressed by

$$V(x) = -\frac{1}{4}(g_\uparrow - A_{21})|x|^2 + \frac{1}{4}\zeta|x|^4, \tag{2.104}$$

and is shown in Fig. 2.15 for the two cases of pumping below threshold $g_\uparrow < A_{21}$ and pumping above threshold $g_\uparrow > A_{21}$.

As the electric field is a complex variable, consisting of amplitude and phase, the potential has a rotationally symmetric form. In the case of pumping below threshold $g_\uparrow < A_{21}$, only one minimum exists and the laser field solution, expressed by the "particle" in the potential, corresponds to this minimum location, i.e. to an average electric field amplitude of zero. The external forces, i.e. the photons spontaneously emitted into the laser mode, will cause fluctuations out of this minimum. However, the repulsive force of the potential always drives the system back to its minimum solution. In the case of pumping above threshold $g_\uparrow > A_{21}$, the potential form changes and a ring-shaped valley results. The particle in this valley now corresponds to a non-zero field amplitude, i.e. a nearly constant laser intensity. The external forces then cause only slight fluctuations of this amplitude, showing that the laser exhibits an amplitude stability. However, they cause a diffusion-like motion along the valley, which corresponds to the **phase diffusion** of a laser that results from the spontaneous and thus non-coherent emissions, which add to the laser field. These phase fluctuations are responsible for slight frequency shifts, and therefore, determine the line width of the laser emission. As can be shown from laser theory, the average amplitude fluctuations, as well as the laser line width (see Sect. 3.3) caused by the phase fluctuations, decrease with the inverse laser intensity. This makes the laser a highly coherent light source.

References

1. H. Haken, H.C. Wolf, *Atom- und Quantenphysik*, 7th edn. (Springer, Berlin, 2000), p. 59ff
2. I.T. Sorokina, in *Solid-State Mid-Infrared Laser Sources*, ed. by I.T. Sorokina, K.L. Vodopyanov. Topics in Applied Physics, vol. 89 (Springer, Berlin, 2003), pp. 255–349
3. G.H. Dieke, *Spectra and Energy Levels of Rare Earth Ions in Crystals* (Wiley Interscience, New York, 1968)
4. F. Träger, *Handbook of Lasers and Optics* (Springer, New York, 2007)
5. D.E. McCumber, Phys. Rev. **136**, A945 (1964)
6. H.W. Moos, J. Lumin. **1**, 106 (1970)
7. M. Eichhorn, Quasi-three-level solid-state lasers in the near and mid-infrared based on trivalent rare earth ions. Appl. Phys. B **93**, 269–316 (2008)

Chapter 3
Optical Resonators

In this chapter, we investigate optical resonators that are used in a laser to provide feedback to the gain medium and to control the oscillating modes and beam properties of the laser output. The most common type of optical resonators are the linear resonators, which in their simplest form consist of two mirrors on the same optical axis with surfaces that are parallel and aligned, with respect to each other, and perpendicular to this optical axis.

Such a resonator exhibits eigen-solutions for the electromagnetic field, which are called **resonator modes**. As the mirror surfaces do not extend into infinity, diffraction occurs for the resonator modes and a curved mirror surface will be necessary in order to refocus the diffracted beams into the resonator volume. Therefore, we will discuss linear resonators with spherical mirrors in Sect. 3.1 and their mode properties in Sect. 3.2.

3.1 Linear and Ring Resonators and Their Stability Criteria

An elegant and simple description of the stability properties of optical resonators, i.e. the question whether eigen-solutions exist or not, can be achieved through use of simple geometrical optics, using a matrix formalism presented in the following section.

3.1.1 Basics of Matrix Optics

In geometrical optics electromagnetic radiation is represented by a light ray. In an axially symmetric problem, this ray can be described by its distance from the optical axis $r(z)$ and the local slope $r'(z)$ at the axial position z, see Fig. 3.1, and may thus be described by a vector

$$\vec{r}(z) = \begin{pmatrix} r(z) \\ r'(z) \end{pmatrix} = \begin{pmatrix} r(z) \\ \tan \alpha(z) \end{pmatrix} \approx \begin{pmatrix} r(z) \\ \alpha(z) \end{pmatrix}, \tag{3.1}$$

M. Eichhorn, *Laser Physics*, Graduate Texts in Physics,
DOI 10.1007/978-3-319-05128-4_3,
© Springer International Publishing Switzerland 2014

Fig. 3.1 Propagation of a
paraxial beam in vacuum

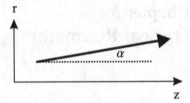

Fig. 3.2 Propagation of a
paraxial beam in vacuum

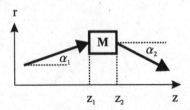

where, the paraxial approximation was used, stating that the slope of the beam
is small enough so that the "small angle" trigonometric functions $\sin\alpha \approx \alpha$ and
$\tan\alpha \approx \alpha$ of the propagation angle α may approximated as shown.

Using this definition for the optical ray, an optical system can be described by a
matrix $\mathbf{M}$ that transforms the incident beam into an exit beam according to [1]

$$\vec{r}_2(z_2) = \mathbf{M}\vec{r}_1(z_1) = \begin{pmatrix} M_{11} & M_{12} \\ M_{21} & M_{22} \end{pmatrix} \begin{pmatrix} r(z_1) \\ \alpha(z_1) \end{pmatrix}, \tag{3.2}$$

as can be seen in Fig. 3.2. The ray matrices of different optical elements can be found
in Table 3.1. Any optical system that consists of N of these elements, including the
free space between the elements, can therefore be described by one single matrix
$\mathbf{M}_S$, which relates the input plane and the output plane of this system. This matrix
is then given by

$$\mathbf{M}_S = \mathbf{M}_N \cdot \mathbf{M}_{N-1} \cdot \ldots \cdot \mathbf{M}_1 = \overset{\overleftarrow{N}}{\underset{i=1}{\prod}} \mathbf{M}_i. \tag{3.3}$$

It should be noted that it is important to multiply the matrices in reverse order with
respect to the beam propagation direction.

3.1.2 Stable and Unstable Linear Resonators

An optical two-mirror resonator with the mirror curvatures R_1 and R_2 and a mirror
separation L as shown in Fig. 3.3 can be represented by an infinite succession of
two lenses with focal lengths $f_1 = \frac{R_1}{2}$ and $f_2 = \frac{R_2}{2}$. In order to deduce the stability
criteria for an optical resonator we investigate one round-trip inside this resonator
and its equivalent path in the infinite-lenses representation in Fig. 3.3. In this representation the resonator can be described by the fundamental element shown in

Table 3.1 Ray matrices for paraxial optical elements

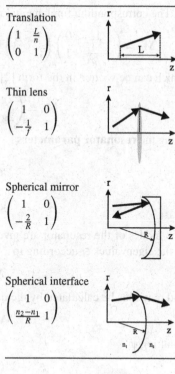

Translation

$$\begin{pmatrix} 1 & \frac{L}{n} \\ 0 & 1 \end{pmatrix}$$

Thin lens

$$\begin{pmatrix} 1 & 0 \\ -\frac{1}{f} & 1 \end{pmatrix}$$

Spherical mirror

$$\begin{pmatrix} 1 & 0 \\ -\frac{2}{R} & 1 \end{pmatrix}$$

Spherical interface

$$\begin{pmatrix} 1 & 0 \\ \frac{n_2 - n_1}{R} & 1 \end{pmatrix}$$

Fig. 3.3 Two-mirror resonator and the comparable infinite-lenses representation

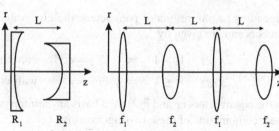

Fig. 3.4 Fundamental element of the periodic infinite-lenses representation of a resonator

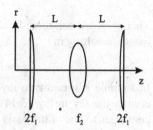

Fig. 3.4 consistent with this round-trip of a light ray within the cavity. It is formed by a first half-lens, followed by the lens f_2 after a distance L corresponding to the reflection on mirror R_2 and a second half-lens which is separated by the lens f_2 by a distance L.

The corresponding fundamental matrix $\mathbf{M}_0$ of this element is thus given by

$$\mathbf{M}_0 = \begin{pmatrix} 1 & 0 \\ -\frac{1}{2f_1} & 1 \end{pmatrix} \begin{pmatrix} 1 & L \\ 0 & 1 \end{pmatrix} \begin{pmatrix} 1 & 0 \\ -\frac{1}{f_2} & 1 \end{pmatrix} \begin{pmatrix} 1 & L \\ 0 & 1 \end{pmatrix} \begin{pmatrix} 1 & 0 \\ -\frac{1}{2f_1} & 1 \end{pmatrix}, \qquad (3.4)$$

which can be written in the form [2]

$$\mathbf{M}_0 = \begin{pmatrix} 2g_1g_2 - 1 & 2g_2L \\ 2g_1\frac{g_1g_2-1}{L} & 2g_1g_2 - 1 \end{pmatrix} \qquad (3.5)$$

using the **resonator parameters**

$$g_1 = 1 - \frac{L}{R_1} \qquad (3.6)$$

$$g_2 = 1 - \frac{L}{R_2}. \qquad (3.7)$$

The modes of the resonator are given by the eigenvectors $\vec{r}_i$ of $\mathbf{M}_0$, corresponding to the eigenvalues ξ_i according to

$$\mathbf{M}_0\vec{r}_i = \xi_i\vec{r}_i, \qquad (3.8)$$

and may thus be calculated by the determinant relation

$$|\mathbf{M}_0 - \xi\mathbf{I}| = 0, \qquad (3.9)$$

resulting in

$$\xi^2 - 2\xi(2g_1g_2 - 1) + 1 = 0. \qquad (3.10)$$

Depending on the resonator parameters, the eigenvalues may be real or complex numbers and are given by

$$|2g_1g_2 - 1| > 1 \quad \Rightarrow \quad \xi_{1,2} = e^{\pm p} \text{ with } \cosh p = 2g_1g_2 - 1 \qquad (3.11)$$

$$|2g_1g_2 - 1| \leq 1 \quad \Rightarrow \quad \xi_{1,2} = e^{\pm iq} \text{ with } \cos q = 2g_1g_2 - 1. \qquad (3.12)$$

As the eigenvectors $\vec{r}_1$ and $\vec{r}_2$ form a basis any arbitrary ray $\vec{r}$ can be expressed as a linear combination of these two eigenvectors

$$\vec{r} = a_1\vec{r}_1 + a_2\vec{r}_2. \qquad (3.13)$$

Therefore, we can directly calculate the ray vector after N round-trips in this resonator, resulting in

$$\vec{r}_N = \mathbf{M}_0^N\vec{r} = a_1\xi_1^N\vec{r}_1 + a_2\xi_2^N\vec{r}_2. \qquad (3.14)$$

In a **stable resonator** no ray will leave the resonator, forcing $|\xi_1| = |\xi_2| = 1$. Otherwise, the ray in Eq. (3.14) diverges in radial position and propagation angle or converges to one of the basis vectors or the null vector. Thus, the stability criteria of a two-mirror laser resonator are given by

$$0 \leq g_1g_2 \leq 1 \quad \Rightarrow \quad \text{stable resonator}, \qquad (3.15)$$

$$g_1g_2 < 0 \ \lor \ g_1g_2 > 1 \quad \Rightarrow \quad \text{unstable resonator}. \qquad (3.16)$$

Fig. 3.5 Stable and unstable resonators and the corresponding relative orientation of the curvatures [2]

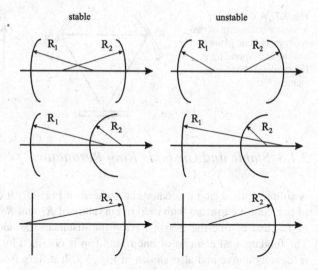

Fig. 3.6 Stability diagram of a two-mirror resonator [2]

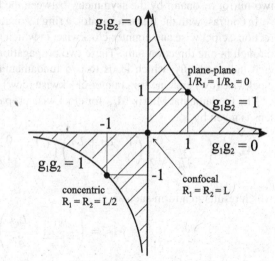

Using Eqs. (3.6), (3.7), a simple geometrical rule can be deduced for these criteria, based on the relative overlap between the curvature radii of the two mirrors as shown in Fig. 3.5:

- A stable resonator corresponds to a partial overlap of the radii;
- An unstable resonator results when both radii do not overlap or when one radius is fully comprised within the other.

A second way to visualize the resonator parameters is the stability diagram in Fig. 3.6, in which the stable and unstable zones, as well as the point of operation of the resonator, are shown.

Fig. 3.7 A simple ring
resonator consisting of two
concave and one plane mirror
and the corresponding
fundamental round-trip
element

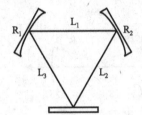 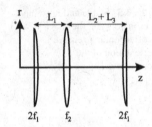

3.1.3 Stable and Unstable Ring Resonators

A simple optical ring resonator can be seen in Fig. 3.7. It consists in this example
of two concave mirrors with radii of curvature of R_1 and R_2 and a mirror separation
of L_1, and of one flat mirror having the distances L_2 and L_3 to the two others.
The fundamental element of one round-trip is obtained by analogy with the linear
resonators above and also shown in Fig. 3.7. It differs from the one of the simple
two-mirror resonator by the asymmetry between the lenses.

In contrast with the linear resonator, a ring resonator exhibits two propagation directions, clockwise and counter-clockwise, in which the optical elements are passed
through in one direction only. These two propagation directions have to be considered independently, which leads to two fundamental round-trip matrices, one for
clockwise (cw) and one for counter-clockwise (ccw) propagation.

The fundamental Matrix $\mathbf{M}_{cw}$ for clockwise propagation starting at the first half-
lens is given by

$$\mathbf{M}_{cw} = \begin{pmatrix} 1 & 0 \\ -\frac{1}{2f_1} & 1 \end{pmatrix} \begin{pmatrix} 1 & L_2+L_3 \\ 0 & 1 \end{pmatrix} \begin{pmatrix} 1 & 0 \\ -\frac{1}{f_2} & 1 \end{pmatrix} \begin{pmatrix} 1 & L_1 \\ 0 & 1 \end{pmatrix} \begin{pmatrix} 1 & 0 \\ -\frac{1}{2f_1} & 1 \end{pmatrix},$$

(3.17)

which results in a form such as

$$\mathbf{M}_{cw} = \begin{pmatrix} g_1 & L_{eff} \\ \frac{g_1 g_2 - 1}{L_{eff}} & g_2 \end{pmatrix},$$

(3.18)

defining the **resonator parameters** g_1 and g_2 of a ring resonator [4]. As a result of
the usually higher number of variables in a ring resonator, these resonator parameters often consist of more complex expressions. Performing the calculation for the
above example yields

$$g_1 = 1 - \frac{L_{eff}}{R_1} - \frac{L_2 + L_3}{R_2}$$

(3.19)

$$g_2 = 1 - \frac{L_{eff}}{R_1} - \frac{L_2 + L_3}{R_2} + 2\frac{L_2 + L_3 - L_1}{R_2},$$

(3.20)

where, L_{eff} characterizes the effective resonator length

$$L_{eff} = L_2 + L_3 + L_1 \left(1 - 2\frac{L_2 + L_3}{R_2}\right).$$

(3.21)

The corresponding fundamental matrix $\mathbf{M}_{ccw}$ for a counter-clockwise propagation starting at the first half-lens is obtained by analogy as

$$\mathbf{M}_{ccw} = \begin{pmatrix} g_2 & L_{eff} \\ \frac{g_1 g_2 - 1}{L_{eff}} & g_1 \end{pmatrix}. \tag{3.22}$$

Hence, the two matrices of the two propagation directions only differ by an exchange in the resonator parameters [4].

The eigenvalues ξ_i of the modes of this resonator, here for clockwise propagation, are again obtained from the relation for the eigenvectors $\vec{r}_i$ of $\mathbf{M}_{cw}$

$$\mathbf{M}_{cw}\vec{r}_i = \xi_i \vec{r}_i \tag{3.23}$$

solved by the determinant formula

$$|\mathbf{M}_{cw} - \xi \mathbf{I}| = 0. \tag{3.24}$$

This results in the characteristic eigenvalue equation

$$\xi^2 - (g_1 + g_2)\xi + 1 = 0. \tag{3.25}$$

Owing to the symmetry of this equation under exchange of g_1 and g_2, an identical characteristic eigenvalue equation results for the counter-clockwise propagation. Hence, the eigenvalues for both propagation directions are identical. Depending on the resonator parameters, the eigenvalues can again be real or complex numbers and result in

$$\left| \frac{g_1 + g_2}{2} \right| > 1 \quad \Rightarrow \quad \xi_{1,2} = e^{\pm p} \quad \text{with } \cosh p = \frac{g_1 + g_2}{2}, \tag{3.26}$$

$$\left| \frac{g_1 + g_2}{2} \right| \le 1 \quad \Rightarrow \quad \xi_{1,2} = e^{\pm iq} \quad \text{with } \cos q = \frac{g_1 + g_2}{2}. \tag{3.27}$$

By analogy with the linear resonator description, the stability criterion of a ring resonator results in

$$|g_1 + g_2| \le 2 \quad \Rightarrow \quad \text{stable resonator} \tag{3.28}$$

$$|g_1 + g_2| > 2 \quad \Rightarrow \quad \text{unstable resonator,} \tag{3.29}$$

and a corresponding stability diagram can be drawn, as shown in Fig. 3.8.

3.2 Mode Structure and Intensity Distribution

In the previous Sect. 3.1.2 we investigated the stability of optical resonators using geometric optics based on ray matrices. This allows the determination of the stability criteria of a resonator, but does not give an explanation on the distribution of the modes in the resonator. To deduce this, we have to use the wave description of the electromagnetic field of the eigenmode of the resonator using the scalar-field approximation. This approximation is valid for resonators with large dimensions

Fig. 3.8 Stability diagram of
a ring resonator [4]

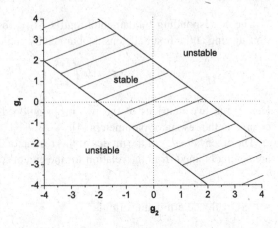

compared with the wavelength, i.e. with a resonator length $L \gg \lambda$ and a mirror
diameter $2a \gg \lambda$, and for linearly polarized fields that are perpendicularly oriented
with respect to the propagation axis. Therefore, these modes are called **transverse
electromagnetic modes** or **TEM-modes**. Then, **Huygens' principle** can be used to
calculate the propagation of the electromagnetic fields and to deduce the diffraction
losses of the resonator and its intensity distribution.

3.2.1 The Fundamental Mode: The Gaussian Beam

In order to derive the properties of the fundamental mode of a resonator we start
from the wave equation

$$\nabla^2 \vec{E} = \frac{1}{c^2} \frac{\partial^2 \vec{E}}{\partial t^2}. \tag{3.30}$$

As a mode can be described by a single frequency ω, we can assume an electric field
of the form

$$\vec{E} = E_0(x, y, z) \vec{\epsilon} e^{i\omega t}, \tag{3.31}$$

which is assumed to be linearly polarized along the polarization vector $\vec{\epsilon}$. Inserting
this into Eq. 3.30 results in the scalar wave equation

$$\nabla^2 E_0 + k^2 E_0 = 0, \tag{3.32}$$

with $k = \frac{\omega}{c}$. Additionally we assume that the wave is propagating along the z-axis
so that we can write

$$E_0(x, y, z) = \hat{E}_0 \psi(x, y, z) e^{ikz}, \tag{3.33}$$

where, $\hat{E}_0$ is the maximum field amplitude and $\psi(x, y, z)$ describes the transverse
field distribution and is assumed to depend only weakly on z. Therefore, we can

neglect the term $\frac{\partial^2 \psi}{\partial z^2}$ in the scalar wave equation, resulting in the paraxial wave equation

$$\Delta_T \psi(x, y, z) - 2ik\frac{\partial \psi}{\partial z} = 0, \tag{3.34}$$

where, $\Delta_T = \frac{\partial^2}{\partial x^2} + \frac{\partial^2}{\partial y^2}$ denotes the transverse Laplacian operator.

To solve the paraxial wave equation we use the ansatz

$$\psi(x, y, z) = e^{i(p + \frac{kr^2}{2q})} \tag{3.35}$$

with $r^2 = x^2 + y^2$, resulting in

$$\frac{\partial p}{\partial z} = -\frac{i}{q} \tag{3.36}$$

$$\frac{\partial q}{\partial z} = 1. \tag{3.37}$$

Thus, $q(z)$ can be seen as the **complex beam radius**. Propagating from the axial position z_1 to z_2 the complex beam radius evolves according to

$$q(z_2) = q(z_1) + z_2 - z_1. \tag{3.38}$$

This is the fundamental equation that allows to calculate the beam parameters during propagation. For the complex beam radius we define the real variables R and w as

$$\frac{1}{q} = \frac{1}{R} - i\frac{\lambda}{\pi w^2}, \tag{3.39}$$

resulting in

$$\psi(r, z) = e^{-ip} e^{-ik\frac{r^2}{2R}} e^{-\frac{r^2}{w^2}}. \tag{3.40}$$

Thus, we can deduce that the electric field has a Gaussian transverse distribution with a $\frac{1}{e}$ field radius w. The term $e^{-ik\frac{r^2}{2R}}$ describes the transverse phase distribution. Starting from a real spherical wave with an origin at $z = 0$ the corresponding phase factor at $z = R$ would be

$$e^{-ik\sqrt{x^2+y^2+R^2}} \approx e^{-ikR} e^{-ik\frac{r^2}{2R}}. \tag{3.41}$$

This shows that the fundamental mode exhibits spherical phase fronts near the z-axis and $R(z)$ can be seen as the local phase front curvature radius.

In the plane where the Gaussian beam has a flat phase front, i.e. $R \rightarrow \infty$, the complex beam radius simplifies to

$$q_0 = i\frac{\pi w_0^2}{\lambda} = iz_R, \tag{3.42}$$

defining the **Rayleigh range** z_R. Assuming that this occurs at $z = 0$ we can write the propagation of the complex beam radius as

$$q(z) = q_0 + z = z + iz_R \tag{3.43}$$

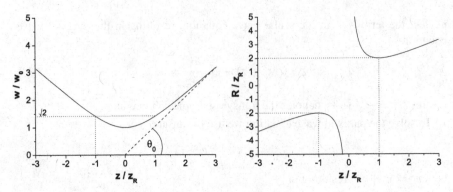

Fig. 3.9 Beam radius and phase front radius of a Gaussian beam

and can then deduce the relations for the real beam radius and the phase front curvature as

$$w(z) = w_0 \sqrt{1 + \left(\frac{z}{z_R}\right)^2} \tag{3.44}$$

$$R(z) = z + \frac{z_R^2}{z}. \tag{3.45}$$

The evolution of these two parameters is given in Fig. 3.9, showing that the beam diverges with the distance slowly from the focus up to the Rayleigh range z_R. For larger distances from the focus the divergence increases and finally results in a beam radius that increases linearly with a divergence angle of

$$\theta(z) = \arctan \frac{\lambda}{\pi w_0} \approx \frac{\lambda}{\pi w_0}. \tag{3.46}$$

At the Rayleigh range, the beam radius increased by a factor of $\sqrt{2}$. The absolute radius of curvature of the phase front quickly decreases with separation from the focus and shows a minimum of $R(\pm z_R) = 2z_R$ at $|z| = z_R$ and then starts increasing again with a linear increase for larger distances as $R(z) \approx z$. For large distances from the focus, the beam shows a phase front that is comparable with a spherical wave emitted from the origin.

Finally, we need to integrate Eq. (3.36), resulting in

$$p(z) = -i \ln w(z) - \arctan \left(\frac{z}{z_R}\right). \tag{3.47}$$

Thus, the last missing term in the field distribution can be solved. The full field distribution of a Gaussian beam thus can be given by

$$\psi(r, z) = \frac{w_0}{w(z)} e^{-i\phi(z)} e^{-ik\frac{r^2}{2R(z)}} e^{-\frac{r^2}{w(z)^2}}, \tag{3.48}$$

with $\phi(z) = \arctan\left(\frac{z}{z_R}\right)$, called the **Gouy phase shift**. Consequently, a phase shift of π occurs when the beam crosses its focal region.

Fig. 3.10 Position of
a Gaussian beam inside
a spherical-mirror resonator

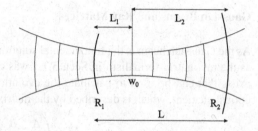

The Gaussian beam is a solution of the free-space paraxial wave equation and up to now no resonator parameters have been used to obtain a solution. However, resulting from the fact that the phase front curvature of the Gaussian beam is spherical, we can insert a spherical mirror with a radius of curvature of $R(z_1)$ into a Gaussian beam at the position z_1 and the beam will be retro-reflected exactly into itself. By inserting a second mirror with a radius of curvature of $R(z_2)$ at the position z_2, we created an optical resonator that is self-consistent with this Gaussian beam, which therefore can be seen as the fundamental mode of this resonator. We now only need to find a way to reverse this scheme, i.e. to find the Gaussian beam for a given resonator, i.e. for given values of R_1, R_2 and L.

Referring to Fig. 3.10 and using Eq. (3.45) we find the relations

$$R_1 = L_1 + \frac{z_R^2}{L_1} \tag{3.49}$$

$$R_2 = L_2 + \frac{z_R^2}{L_2} \tag{3.50}$$

$$L = L_1 + L_2. \tag{3.51}$$

This set of equations can be easily solved, resulting in

$$L_1 = g_2(1 - g_1) \frac{L}{g_1 + g_2 - 2g_1g_2} \tag{3.52}$$

$$L_2 = g_1(1 - g_2) \frac{L}{g_1 + g_2 - 2g_1g_2} \tag{3.53}$$

$$z_R^2 = g_1g_2(1 - g_1g_2) \left(\frac{L}{g_1 + g_2 - 2g_1g_2} \right)^2. \tag{3.54}$$

Thus, the position of the focus with respect to the two mirrors, as well as the corresponding Rayleigh range, can directly be calculated. However, from $z_R^2 > 0$ it follows that $0 \leq g_1g_2 \leq 1$ in order to yield a real result for the Rayleigh range. This is again the stability criterion that resulted from the matrix formalism in Sect. 3.1.2. Thus, we have found that the resonators that can be matched to a Gaussian beam are simply the stable resonators; hence, we can conclude that the stable resonators have a Gaussian beam as their fundamental mode. For an unstable resonator, no Gaussian beam can be found that self-consistently matches this resonator and thus no Gaussian eigenmode exists.

Gaussian Beams and Ray Matrices

As the Gaussian beam is the fundamental solution to the paraxial wave equation and as the ray matrix formalism in Sect. 3.1.1 was created for paraxial rays, a relation exists that can be used to calculate the evolution of a Gaussian beam passing an optical element, which is described by the matrix

$$\mathbf{M} = \begin{pmatrix} A & B \\ C & D \end{pmatrix}. \tag{3.55}$$

This relation states that the complex beam radius q_2 at the exit plane of the optical system is connected to the complex beam radius q_1 at the entrance plane by [3]

$$q_2 = \frac{Aq_1 + B}{Cq_1 + D}. \tag{3.56}$$

Owing to their form the ray matrices are often also called **ABCD-matrices**. As an example, the transformation of a Gaussian beam by a thin lens with a focal length f is investigated. Using Table 3.1 the entrance and exit complex beam radii are thus related by

$$\frac{1}{q_2} = \frac{1}{q_1} - \frac{1}{f}. \tag{3.57}$$

In a similar way, Eq. (3.56) allows to find the fundamental mode properties of complex laser resonators. In order to do so, an arbitrary point O is chosen on the resonator axis inside the resonator, e.g. in front of the outcoupling mirror. From this point O, on a beam propagating along the resonator axis, it encounters the output coupler where a part will be reflected. This reflected beam then travels along all resonator elements (mirrors, lenses, laser media with thermal lens, and other optical elements) in backward propagation, gets reflected by the cavity end mirror and then propagates in forward direction passing all elements again just up to the same point, here in front of the outcoupling mirror. As a stable resonator mode has to reproduce itself during one **round trip**, the complex beam radius at the beginning and at the end of this round trip need to be equal. Thus Eq. (3.56) becomes

$$q = \frac{A_{RT}q + B_{RT}}{C_{RT}q + D_{RT}}, \tag{3.58}$$

where q denotes the (unknown) complex beam radius of the mode at this point O and

$$\mathbf{M}_{RT} = \begin{pmatrix} A_{RT} & B_{RT} \\ C_{RT} & D_{RT} \end{pmatrix}. \tag{3.59}$$

is the round-trip resonator matrix calculated along the path described above starting from the chosen point O within the resonator. The complex beam radius of the mode at this point O is then found by solving

$$Cq^2 + (D - A)q - B = 0. \tag{3.60}$$

3.2.2 Higher-Order Transverse Modes and Beam Quality

The Gaussian beam investigated in the previous chapter is only the lowest-order mode of an infinite class of modes. In the most important case of a cylindrical symmetry of the laser cavity using a confocal resonator, these are described by **Laguerre-Gaussian** functions, with an electric field amplitude of [2]

$$E_{lp}(r, \phi, z) \propto \cos l\phi \frac{(2\rho)^l}{(1 + Z^2)^{\frac{l+1}{2}}} L_p^l\left(\frac{(2\rho)^2}{1 + Z^2}\right) e^{-\frac{\rho^2}{1+Z^2}}$$

$$\times e^{-i\left(\frac{(1+Z)\pi R}{\lambda} + \frac{\rho^2 Z}{1+Z^2} - (l+2p+1)[\frac{\pi}{2} - \arctan(\frac{1-Z}{1+Z})]\right)}, \tag{3.61}$$

with

$$\rho = r\sqrt{\frac{2\pi}{R\lambda}} \tag{3.62}$$

$$Z = \frac{2}{R}z. \tag{3.63}$$

In this description, $l = 0, 1, 2, \ldots$ describes the angular mode number, $p = 0, 1, 2, \ldots$ the radial mode number, R the radius of curvature of the mirrors that are located at a distance of $L = R$ and $L_p^l(x)$ are the Laguerre polynomials that are defined by the equation

$$L_p^l(x) = \frac{1}{p!}x^{-l}e^x\frac{d^p}{dx^p}\left(x^{p+l}e^{-x}\right). \tag{3.64}$$

The first Laguerre polynomials thus are $L_0^0(x) = 1$, $L_1^l(x) = l + 1 - x$ and $L_2^l(x) = \frac{1}{2}(l + 1)(l + 2) - (l + 2)x + \frac{x^2}{2}$. The corresponding intensity distributions $I_{lp} \propto |E_{lp}|^2$ at $z = 0$ are shown in Fig. 3.11, from which we can deduce that p corresponds to the number of radial minima in the intensity distribution while $2l$ gives the number of angular minima on the full 2π angle. When we denote the beam radius of the fundamental TEM$_{00}$ beam with ω_{00}, the equivalent beam radius of the higher order Laguerre-Gaussian beams is given by

$$\omega_{lp} = \omega_{00}\sqrt{2p + l + 1}. \tag{3.65}$$

Whenever the cylindrical symmetry of the laser cavity is broken, e.g. when rectangular mirrors are used with transverse dimensions comparable with the beam size, so that non-symmetric losses occur, the modes may better be described in Cartesian coordinates, resulting in **Hermite-Gaussian** modes. Also tilted optical elements inside the cavity, e.g. Brewster windows, can cause this effect. The Hermite-Gaussian modes show the intensity distribution [2]

$$E_{mn}(r, \phi, z) \propto \frac{1}{\sqrt{1 + Z^2}}H_m\left(X\sqrt{\frac{2}{1 + Z^2}}\right)H_n\left(Y\sqrt{\frac{2}{1 + Z^2}}\right)e^{-\frac{X^2+Y^2}{1+Z^2}}$$

$$\times e^{-i\left(\frac{(1+Z)\pi R}{\lambda} + \frac{(X^2+Y^2)Z}{1+Z^2} - (m+n+1)[\frac{\pi}{2} - \arctan(\frac{1-Z}{1+Z})]\right)}, \tag{3.66}$$

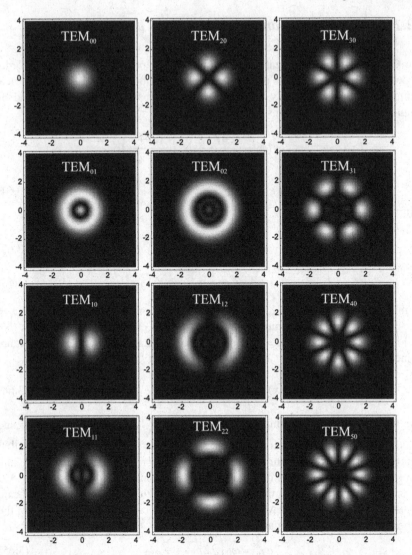

Fig. 3.11 Intensity distributions of some Laguerre-Gaussian beams (TEM$_{lp}$ modes)

with

$$X = x\sqrt{\frac{2\pi}{R\lambda}} \qquad (3.67)$$

$$Y = y\sqrt{\frac{2\pi}{R\lambda}} \qquad (3.68)$$

$$Z = \frac{2}{R}z. \qquad (3.69)$$

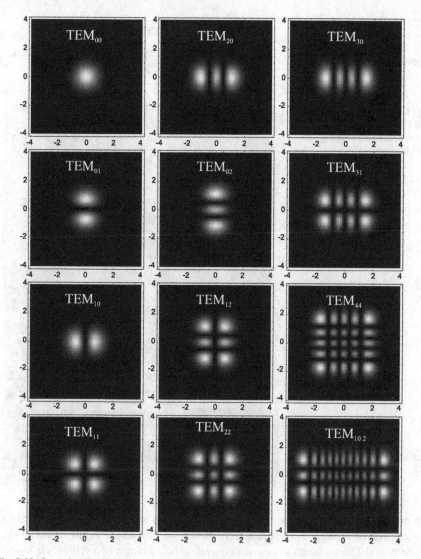

Fig. 3.12 Intensity distributions of some Hermite-Gaussian beams (TEM$_{mn}$ modes)

In this description, $m = 0, 1, 2, \ldots$ and $n = 0, 1, 2, \ldots$ describe mode numbers corresponding to the x- and y-axis, respectively and $H_m(x)$ are the Hermite polynomials that are defined by the equation

$$H_m(x) = (-1)^m e^{x^2} \frac{d^m}{dx^m}\left(e^{-x^2}\right). \tag{3.70}$$

The first Hermite polynomials thus are $H_0(x) = 1$, $H_1(x) = 2x$, $H_2(x) = 4x^2 - 2$ and $H_3(x) = 8x^3 - 12x$. The corresponding intensity distributions $I_{mn} \propto |E_{mn}|^2$ at $z = 0$ are shown in Fig. 3.12, from which we can deduce that m corresponds to

Fig. 3.13 Brewster setup of a laser cavity used for a Cr^{2+}:ZnSe laser to avoid the necessity of optical coatings on the laser crystal

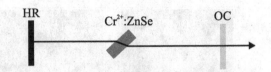

Fig. 3.14 Measured intensity distributions of some Hermite-Gaussian beams (TEM$_{mn}$ modes) of a Cr^{2+}:ZnSe laser at 2.3 μm [5]

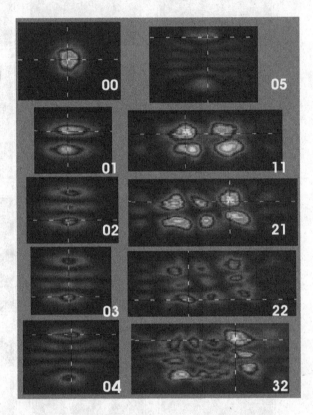

the number of minima in the intensity distribution along the x-axis while n gives the number of minima along the y-axis. When we denote the beam radius of the fundamental TEM$_{00}$ beam as ω_{00}, the equivalent beam radius of a symmetric higher order Hermite-Gaussian beam, i.e. with $m = n$, is given by

$$\omega_{mm} = \omega_{00}\sqrt{2m + 1}. \tag{3.71}$$

In a laser setup the laser crystal is sometimes inserted into the resonator under Brewster's angle as shown in Fig. 3.13. This provides a linearly polarized laser output directly, as only one polarization orientation passes the crystal with low Fresnel reflection losses. It is also sometimes used to test crystal samples quickly in laser operation, as it does not require the anti-reflection coating of the crystal, which is usually used when the crystals are inserted under normal incidence. When the laser medium is placed at Brewster's angle in the resonator, this also causes a symmetry breaking responsible for the occurrence of the Hermite-Gaussian modes. Figure 3.14

shows the measured intensity distributions of some Hermite-Gaussian modes of a Brewster-oriented Cr^{2+}:ZnSe laser [5].

Beam Quality

The TEM_{00} fundamental mode is the same in both geometries and corresponds to the Gaussian beam discussed in Sect. 3.2.1. As will be shown in the following discussion, the Gaussian beam also shows the smallest divergence angle and the lowest focal spot size of all modes, and therefore, has the best beam quality. In construction of a laser it is thus usually the aim to get TEM_{00} operation.

In order to deduce a quantity that describes the beam quality of a laser beam along a certain transverse axis, e.g. the x-axis, we investigate the different statistical moments of the field amplitude $\psi(x)$ [3]. Therefore, we define the statistical average

$$\langle f(x) \rangle = \frac{\int_{-\infty}^{\infty} f(x)|\psi(x)|^2 dx}{\int_{-\infty}^{\infty} |\psi(x)|^2 dx}. \tag{3.72}$$

The beam radius w_x and the radius of curvature R_x can then be written as

$$w_x = 2\sqrt{\langle x^2 \rangle - \langle x \rangle^2} \tag{3.73}$$

$$\frac{1}{R_x} = \frac{i\lambda}{\pi w_x^2 \int_{-\infty}^{\infty} |\psi(x)|^2 dx}$$

$$\times \int_{-\infty}^{\infty} \left(\psi^*(x)\frac{\partial \psi(x)}{\partial x} - \psi(x)\frac{\partial \psi^*(x)}{\partial x} \right)(x - \langle x \rangle) dx. \tag{3.74}$$

The divergence θ_x of the beam along the x-axis, which describes the angular spreading of the energy can be described by the Fourier transform of the amplitude distribution and is given by

$$\theta_x = 2\lambda\sqrt{\langle \xi^2 \rangle - \langle \xi \rangle^2}, \tag{3.75}$$

where, $\langle F(\xi) \rangle$ denotes the moments in Fourier space, i.e.

$$\langle F(\xi) \rangle = \frac{\int_{-\infty}^{\infty} F(\xi)|\Psi(\xi)|^2 d\xi}{\int_{-\infty}^{\infty} |\Psi(\xi)|^2 d\xi} \tag{3.76}$$

with $\Psi(\xi)$ being the Fourier transform of $\psi(x)$,

$$\Psi(\xi) = \frac{1}{\sqrt{2\pi}} \int_{-\infty}^{\infty} \psi(x)e^{-i2\pi\xi x} dx. \tag{3.77}$$

It can now be shown that the expression

$$M^2 = \frac{\pi}{\lambda}w_x\sqrt{\theta_x^2 - \frac{w_x^2}{R_x^2}} \tag{3.78}$$

is an invariant property of the beam. That means that passive optical systems, which can be described by ray matrices, e.g. lenses or spherical mirrors, do not influence this value for a given beam. It is interesting to evaluate this expression for a collimated beam, i.e. at the position where $R_x \to \infty$. The divergence is then given by

$$\theta_x = \frac{M^2 \lambda}{\pi w_x} = M^2 \theta_0, \tag{3.79}$$

showing that the divergence of the real beam is M^2 times stronger than the divergence of a collimated Gaussian beam of same radius, see Eq. (3.46). Owing to $w_x > 0$ it follows that M^2 is a positive quantity. It can also be deduced that a Gaussian beam corresponds to $M^2 = 1$ and that any other field distribution that deviates from the Gaussian beam in amplitude or phase will result in an $M^2 > 1$. Thus, M^2 is referred to as the **beam quality factor**. The optimum beam quality corresponds to $M^2 = 1$, i.e. the Gaussian beam has optimum beam quality. An $M^2 > 1$ will result in a stronger divergence of the beam, and thus, in the reverse direction, a given aperture and focal length of a lens results in a larger focal spot size than for a Gaussian beam with equal diameter. When both beams are focused with the same divergence angle, the focal spot of the general beam will be M^2 times larger than the spot diameter of a Gaussian beam. As a result of the influence of M^2 on the divergence it is sometimes also called **beam propagation factor**.

In formulae such as Eq. (3.79), the M^2 always occurs together with the laser beam wavelength λ. This expresses that the M^2 determines the beam divergence property in the same way as the wavelength. A beam propagation calculation, e.g. collimation or focusing, of a beam with $M^2 > 1$, can therefore, be performed by the standard ray matrix formalism, using a hypothetic Gaussian beam with the wavelength

$$\lambda' = M^2 \lambda. \tag{3.80}$$

As the spatial moments (the beam radius w_x) and the spatial frequency moments θ_x are related by a Fourier transform, i.e. x and ξ are so-called **conjugate variables**, we find by rewriting Eq. (3.79) in the form

$$\theta_x w_x = \frac{M^2 \lambda}{\pi} = \text{constant} \tag{3.81}$$

hence, aperture and divergence are related to each other in a form that may remind us on the uncertainty relation Eq. (1.59). In this case, the conjugated variables were time and energy. From this we can deduce that for any given collimated beam its aperture-divergence product is constant and cannot be changed by any kind of passive optical element that may be described by a ray matrix. Whenever an optical element is used that causes aberrations this product, and thus the M^2 of the beam, will increase after passing this element.

Transverse Mode Selection

Depending on the application of a laser it is often important to get single transverse mode operation, in most cases on the TEM_{00} mode. Therefore, mode dependent

Fig. 3.15 The Fresnel
number as the ratio of
reflected and lost power at a
resonator mirror

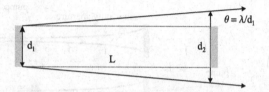

Fig. 3.16 Diffraction losses
as a function of the Fresnel
number for the TEM_{00} and
TEM_{10} modes in Fabry-Perot
and confocal resonators [2]

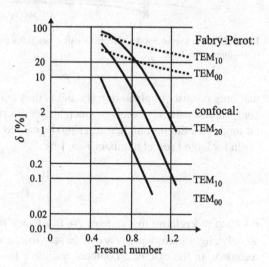

losses need to be included into the laser cavity, forcing the unwanted modes to suffer from a higher loss so that they cannot reach threshold or not be sustained.

A good quantity to describe the diffraction losses of a resonator with circular mirrors of diameter $2a$ and distance L is the **Fresnel number**

$$F = \frac{a^2}{\lambda L},$$ (3.82)

which can be interpreted as the ratio between the reflected power of a cavity mirror to the diffraction loss at that mirror [2]. This can be seen from the following calculation, assuming mirrors with equal diameters as shown in Fig. 3.15. Starting from the first cavity mirror the reflected light beam has a diameter $d_1 = 2a$, and will therefore, suffer from a diffraction that causes the beam to expand with an angle $\theta \approx \frac{\lambda}{d_1}$. Assuming that the beam homogeneously expands towards the second cavity mirror at the distance L results in a beam diameter of $d_2 = 2(a + L\theta)$ at the position of the second cavity mirror. Therefore, the ratio between the power reflected from that second mirror and the losses gives

$$\frac{\pi a^2}{\pi(a + L\theta)^2} \approx \frac{a}{2L\theta} = \frac{a^2}{\lambda L} = F.$$ (3.83)

High Fresnel numbers thus correspond to low loss resonators.

The simplest way of mode selection can already be done by choosing a proper cavity design, e.g. a confocal resonator. According to Fig. 3.16 this cavity will not only show lower total losses compared with the **Fabry-Perot** cavity, i.e. to a cavity

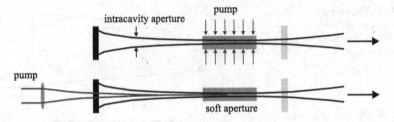

Fig. 3.17 Transverse mode selection using an intracavity aperture or a soft aperture in a quasi-three-level laser

that only consists of plane mirrors and is thus very sensitive to misalignment. The confocal resonator also shows a much higher discrimination between the fundamental mode and the first higher-order mode, caused by the diffraction losses that are given for large Fresnel numbers $F > 1$ by

$$\delta_{00} \approx 5 \times 10^{-4} F^{-7.67} \tag{3.84}$$

$$\delta_{10} \approx 1 \times 10^{-2} F^{-7.67}, \tag{3.85}$$

resulting in a relative loss ratio of 20. In a Fabry-Perot cavity this ratio is only 2.53, so only slight perturbations are necessary to cause mode changes in a Fabry-Perot resonator. In the case of a confocal resonator based on quadratic mirrors of width $2a$, i.e. one with Hermite-Gaussian modes, these losses are given by

$$\delta_{00} \approx 1 \times 10^{-4} F^{-13.3} \tag{3.86}$$

$$\delta_{10} \approx 4 \times 10^{-3} F^{-13.3}, \tag{3.87}$$

resulting in an even higher relative loss ratio of 40.

If this does not provide enough loss discrimination between the modes to guarantee TEM$_{00}$ operation, an aperture may be introduced into the cavity, e.g. a diaphragm at an inner focus or at a position where the difference between the fundamental mode diameter and the diameter of the next higher-order mode is large (**hard aperture**). As a result of the higher beam radii of the higher order modes compared with the fundamental mode, as shown in Eqs. (3.65), (3.71), this aperture will provide a higher loss for higher-order modes, forcing the laser to operate on the fundamental mode. In a quasi-three-level laser medium, this aperture may also be through the pumped volume itself, which is called **soft aperture**. In this case, the pump beam is chosen to be slightly smaller than the beam diameter of the fundamental mode in the laser medium as shown in Fig. 3.17. Then the mode extends into the outer non-pumped parts of the laser medium and consequently suffers from slight reabsorption. Higher-order modes will extend further out into the non-pumped regions and therefore exhibit stronger losses. However, as in any of these cases of mode selection, losses need to be introduced into the laser, mode selection usually decreases the laser output power compared with an unselected multi-mode operation of the same laser.

3.2.3 Longitudinal Modes and Hole-Burning Effects

From the fields in the Eqs. (3.61), (3.66) the phase term can be used to derive the resonance conditions of the modes, i.e. the exact frequency of the mode of the confocal resonator. Thus, we find

$$\frac{2L}{\lambda_{lpq}} = \frac{2L}{c} \nu_{lpq} = q + \frac{1}{2}(2p + l + 1) \tag{3.88}$$

$$\frac{2L}{\lambda_{mnq}} = \frac{2L}{c} \nu_{mnq} = q + \frac{1}{2}(m + n + 1) \tag{3.89}$$

for the Laguerre-Gaussian and the Hermite-Gaussian modes, respectively. In the case of a general resonator with spherical mirrors the corresponding relation [2] is

$$\frac{2L}{\lambda_{lpq}} = \frac{2L}{c} \nu_{lpq} = q + (2p + l + 1)\frac{\arccos \sqrt{g_1 g_2}}{\pi} \tag{3.90}$$

$$\frac{2L}{\lambda_{mnq}} = \frac{2L}{c} \nu_{mnq} = q + (m + n + 1)\frac{\arccos \sqrt{g_1 g_2}}{\pi}, \tag{3.91}$$

where, q is the longitudinal mode index. In the case of a plane wave resonating in a Fabry-Perot cavity, q would correspond to the number of half-cycles of the wave along the cavity, i.e. $L = q\frac{\lambda}{2}$. For the fundamental mode this results in a frequency spacing of the modes

$$\Delta \nu_{00} = \frac{c}{2nL} \tag{3.92}$$

for both cases, where we included the case of a resonator filled with an optical medium of refractive index n. This quantity $\Delta \nu_{00}$ is also referred to as the **free spectral range** of the cavity. For the first higher-order mode the frequency spacing results in

$$\Delta \nu_{10} = \frac{c}{2nL}\frac{\arccos \sqrt{g_1 g_2}}{\pi}, \tag{3.93}$$

showing that these modes are usually non-degenerate with the fundamental mode. Only when one mirror obeys the confocal condition $g_i = 0$ all modes are degenerate owing to $g_1 g_2 = 0$.

In a loss-less resonator, the frequency spectrum of the modes, described by Eqs. (3.90), (3.91), would be given by a series of δ-functions,

$$s_{LG}(\nu) = \sum_{lpq} \delta(\nu - \nu_{lpq}) \tag{3.94}$$

$$s_{HG}(\nu) = \sum_{mnq} \delta(\nu - \nu_{mnq}), \tag{3.95}$$

and correspond to an infinite photon lifetime for each mode in the cavity. However, in real resonators losses occur as a result of diffraction, outcoupling or internal absorption within the resonator that will reduce the cavity photon lifetime to a finite

Fig. 3.18 Spatial field and gain distributions in a standing wave resonator. The lowest graph shows a possible second longitudinal mode that can be amplified by the saturated gain distribution

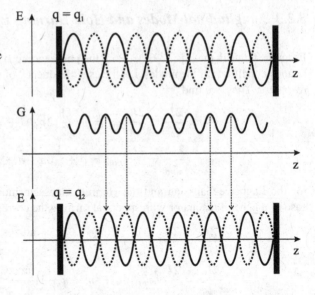

value τ_c. As for the homogeneous line width of an atomic level, which is caused by its natural lifetime, the finite cavity photon lifetime will broaden the resonance of the cavity, resulting in a line width (full width at half maximum) given by

$$\delta\nu = \frac{1}{2\pi\tau_c} = -\frac{c}{4\pi L}\ln\left[R_{OC}(1-\Lambda)R_{HR}\right]. \tag{3.96}$$

Using the mode spacing given in Eq. (3.92), the **finesse** F_c of a cavity can be defined by

$$F_c = \frac{\Delta\nu_{00}}{\delta\nu}, \tag{3.97}$$

describing the sharpness of the resonance of the cavity.

In special applications, in which a very highly monochromatic wave, i.e. a long coherence length of the laser beam is necessary, e.g. in holography or precision measurements, it is important that the laser oscillates not only on the transverse fundamental mode TEM$_{00}$, but also on only one longitudinal mode (single longitudinal mode). However, most lasers will usually oscillate on many longitudinal modes simultaneously. This arises as a result of the effects of **spatial** and **spectral hole burning**.

Spatial Hole Burning

Spatial hole burning always occurs in linear resonators, in which the internal cavity field corresponds to a standing wave. In the nodes of this standing wave the electric field amplitude is always zero, and therefore, the local population inversion will not be saturated as in the field maxima, see Fig. 3.18. Thus, a second longitudinal mode

Fig. 3.19 Spectral gain and cavity modes in a laser resonator with a threshold gain G_{th} for different line broadenings. The solid line describes the gain before laser oscillation starts, the dashed line corresponds to the gain distribution during laser oscillation

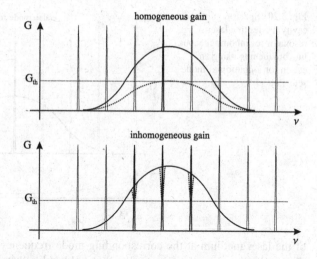

of a different frequency, which has a good overlap between its local maxima and the saturated gain distribution, may have enough gain to oscillate even though the first oscillating mode has already saturated the gain.

To avoid spatial hole burning, ring resonators are used, which consist of at least three mirrors. Using special optical elements, such as **Faraday rotators** or **acousto-optic modulators**, different losses for the two propagation directions can be induced and a unidirectional propagation of the cavity mode results. Consequently, no standing wave will develop and the spatial hole burning is avoided.

Spectral Hole Burning

The second hole burning effect in a laser is spectral hole burning, which occurs whenever the laser medium shows inhomogeneously broadened transitions. For a homogeneous broadening, the mode with the highest gain, will start oscillating first, and therefore, will reduce the gain as shown in Fig. 3.19.

In the case of an inhomogeneously broadened medium, each frequency component of the gain can be saturated independently and every mode starts oscillating for which the gain at the beginning of laser action is higher than the threshold gain G_{th}. Thus, a "hole" is burned into the gain distribution at each frequency of a cavity mode.

Longitudinal-Mode Selection

In order to force the laser to operate on a single longitudinal mode, a frequency selective element needs to be included into the cavity. It increases the losses for all other modes, so that the threshold gain for these modes is bigger than the actual gain

Fig. 3.20 Spectral gain and
cavity modes in a laser
resonator for inhomogeneous
line broadening using an
etalon for longitudinal-mode
selection

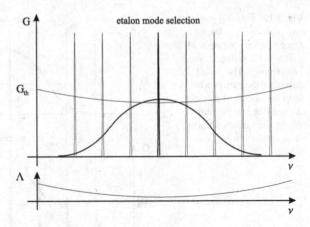

of the laser medium at the corresponding mode frequency, as shown in Fig. 3.20.
Then only the selected mode reaches threshold and will start oscillating.

The frequency selection can by achieved by inserting an **etalon**, i.e. a plane-
parallel glass plate, into the cavity. Depending on the refractive index of the etalon
material and the thickness of the plate, the etalon acts as a small Fabry-Perot resonator. Therefore, it is only loss-free for the frequencies at which nodes occur on the
etalon surfaces. For all other wavelengths Fresnel losses occur that are added to the
cavity losses Λ. The thickness of the etalon is chosen so that when it is tuned to the
central laser frequency, by changing the beam incident angle, the next coincidence
between the modes of the resonator, and the maximum transmission frequencies of
the etalon occur outside the frequency range in which gain can exceed threshold.
Other possibilities are the use of coupled resonators, saturable absorbers or seeding
by a second laser. In the latter case, a single longitudinal mode laser of low power,
e.g. achieved by pumping just above threshold, thus oscillating on the gain peak;
as a result no other mode reaches threshold itself. This beam is injected into the
resonator of a laser dedicated for higher-power, single longitudinal mode operation.
Thus, the second laser locks to the frequency of the seed laser.

3.3 Line Width of the Laser Emission

The theoretical line width of the laser emission was first calculated by Schawlow and
Townes back in 1958, two years before the first laser was experimentally demonstrated. They showed that the theoretical laser line width $\delta\nu_L$ (full width at half
maximum) is given by

$$\delta\nu_L = \frac{2\pi h\nu\Delta\nu_c^2}{P_{out}}, \tag{3.98}$$

with ν being the laser line center frequency, $\Delta\nu_c = \frac{1}{2\pi\tau_c}$ the bandwidth of the passive laser resonator and P_{out} the laser output power.

Here we deduce an equivalent expression of the laser line width that is also valid for quasi-three-level lasers and gives some more insight into the fundamental laser properties. We will use two equivalent descriptions for the Quality-factor (**Q-factor**) of a resonator, which are given by

$$Q = \frac{\nu}{\delta \nu_L} \quad \text{and} \tag{3.99}$$

$$Q = \frac{2\pi E_{tot}}{\Delta E_{osc}} = \frac{2\pi \nu E_{tot}}{\left| \frac{\partial E_{tot}}{\partial t} \right|}. \tag{3.100}$$

The first expression defines the Q-factor as the ratio between the resonance frequency ν and the resonance line width $\delta \nu_L$ (full width at half maximum), whilst the second one uses the energy E_{tot} inside the resonator and the loss of this energy during one period of the oscillation ΔE_{osc}, which can be expressed in terms of the time-averaged energy loss over one period, $\left| \frac{\partial E_{tot}}{\partial t} \right|$. Both descriptions are equivalent, as they are simply related to each other by Fourier transformation between the decay of the intensity (or energy) $I(t)$ of a damped oscillator and its frequency spectrum $\tilde{I}(\nu)$.

We have to take into account that the total energy inside the resonator consists of a coherent part E_c, which is created by stimulated emission and an incoherent part E_{sp} as a result of the spontaneous emission.

During cw laser operation the total internal resonator energy will be constant in time, resulting in

$$E_{tot} = E_c + E_{sp} = \text{const} \quad \Rightarrow \quad \left| \frac{\partial E_c}{\partial t} \right| = \left| \frac{\partial E_{sp}}{\partial t} \right| = P_{sp}. \tag{3.101}$$

Hence, the period-averaged loss of coherent energy is just given by the average power P_{sp} of spontaneously created photons emitted into the laser mode during one oscillation period. The power that is needed to feed the resonator in order to compensate for the losses given by the cavity photon lifetime τ_c can be written as

$$P_{tot} = \frac{E_{tot}}{\tau_c} = \frac{E_c + E_{sp}}{\tau_c} = P_c + P_{sp}, \tag{3.102}$$

where, P_c denotes the coherent power contribution that corresponds to the stimulated emission. From Eq. (2.90) we can deduce that the stimulated emission contribution and the spontaneous contribution are given by

$$P_c = h\nu c \left[\sigma_e(\lambda_s) \langle N_2 \rangle - \sigma_a(\lambda_s) \langle N_1 \rangle \right] \Phi_c V \tag{3.103}$$

$$P_{sp} = h\nu c \sigma_e(\lambda_s) \langle N_2 \rangle \Phi_0 V, \tag{3.104}$$

under the assumption of an axially constant photon density with Φ_c being the density of the coherent photons in the cavity and V being the volume of the cavity. Thus, we can deduce the ratio between the rates of stimulated and spontaneous emission processes as

$$\frac{P_c}{P_{sp}} = \frac{\sigma_e(\lambda_s) \langle N_2 \rangle - \sigma_a(\lambda_s) \langle N_1 \rangle}{\sigma_e(\lambda_s) \langle N_2 \rangle} \frac{\Phi_c}{\Phi_0} = \left(1 - \frac{\sigma_a(\lambda_s) \langle N_1 \rangle}{\sigma_e(\lambda_s) \langle N_2 \rangle} \right) N_P, \tag{3.105}$$

where, we used that the ratio between the coherent photon density Φ_c and the density of the photons of the vacuum fluctuations Φ_0 in the same mode is just given by the number of coherent photons in the mode N_P as the quantum fluctuations correspond to the zero-point fluctuation of that mode, and therefore, to one photon per mode.

As a result, we can write the Q-factor of the oscillating cavity in the form

$$Q = 2\pi \nu \tau_c \frac{P_c + P_{sp}}{P_{sp}} = 2\pi \nu \tau_c \left[\left(1 - \frac{\sigma_a(\lambda_s)\langle N_1 \rangle}{\sigma_e(\lambda_s)\langle N_2 \rangle} \right) N_P + 1 \right]. \quad (3.106)$$

By using the first definition of the Q-factor we can deduce the laser line width as

$$\delta \nu_L = \frac{1}{2\pi \tau_c [(1 - \frac{\sigma_a(\lambda_s)\langle N_1 \rangle}{\sigma_e(\lambda_s)\langle N_2 \rangle}) N_P + 1]} = \frac{\Delta \nu_c}{(1 - \frac{\sigma_a(\lambda_s)\langle N_1 \rangle}{\sigma_e(\lambda_s)\langle N_2 \rangle}) N_P + 1}. \quad (3.107)$$

We may conclude that the laser line width mainly depends on the number of coherent photons in the laser mode, and thus, on the laser output power P_{out} when using

$$N_P + 1 \approx \frac{2}{T_{OC}} \frac{\lambda_s L}{hc^2} P_{out}. \quad (3.108)$$

In the case of low reabsorption $\sigma_a(\lambda_s) \ll \sigma_s(\lambda_s)$, or for a four-level laser, Eq. (3.107) simplifies to

$$\delta \nu_L = \frac{\Delta \nu_c}{N_P + 1}, \quad (3.109)$$

which is equivalent to the Schawlow-Townes relation.

The laser emission line width will always be smaller than the cavity resonance bandwidth and can reach extremely low values below 1 Hz. For a HeNe-laser of 1 mW coherent output power at $\lambda_s = 632.8$ nm with the cavity parameters $L = 0.6$ m and $T_{OC} = 0.02$, we deduce $\tau_c = 198$ ns, i.e. $\Delta \nu_c = 803$ kHz, and a coherent photon number of $N_P = 6.38 \times 10^8$. This results in a theoretical line width of $\delta \nu_L \approx 10^{-3}$ Hz.

Such small line widths, however, will not be found in experimental lasers as all kinds of external fluctuations, such as vibrations on the mirrors, and therefore, on the cavity length, will cause a frequency modulation with a width that is many orders of magnitude larger than the theoretical line width of the laser.

References

1. A.E. Siegman, *Lasers* (University Science Books, Sausalito, 1986)
2. F.K. Kneubühl, M.W. Sigrist, *Laser* (Teubner, Stuttgart, 1999)
3. J. Alda, Laser and Gaussian beam propagation and transformation, in *Encyclopedia of Optical Engineering* (Marcel Dekker, Inc., New York 2003) p. 999. doi:10.1081/E-EOE 120009751
4. Th. Graf, *Laser—Grundlagen der Laserstrahlquellen* (Vieweg+Teubner, Wiesbaden, 2009)
5. M. Eichhorn, Untersuchung eines diodengepumpten Faserverstärkers mit Emission bei 2 μm, Dissertation, Albert-Ludwigs-Universität, Freiburg, Germany, 2005

Chapter 4
Generation of Short and Ultra-Short Pulses

In this chapter we investigate two of the main methods of laser-pulse generation, which are Q-switching and mode-locking. Whilst of-course every laser may be pulsed by just switching it on and off, these methods allow accumulation of pump energy between two pulses, and can therefore, create pulse peak powers that are several orders of magnitude higher than the corresponding cw laser output power.

4.1 Basics of Q-Switching

Q-switching is based on a modulation of the cavity losses, as shown in Fig. 4.1. This modulation, caused by an externally driven intra-cavity modulator in active Q-switching or by a saturable absorber in passive Q-switching, increases the internal losses of the cavity during the pump phase. Thus the laser threshold is dramatically increased and the laser cannot start oscillating, which allows the inversion to reach much higher values than in cw operation. After this pumping phase the modulation losses are switched off and the feedback on the laser medium is restored. Then a laser field builds up from noise and will extract all available stored energy in one giant pulse of high pulse energy. As the loss modulation changes the Q-factor of the cavity, this pulse generation method is called Q-switching. The general temporal evolution of the Q-switch is sketched in Fig. 4.2 for the case of an active Q-switch that acts on the internal cavity losses Λ.

4.1.1 Active Q-Switching

In this section the fundamental properties of actively Q-switched lasers will be deduced, starting from the rate equations (2.60), (2.61).

M. Eichhorn, *Laser Physics*, Graduate Texts in Physics,
DOI 10.1007/978-3-319-05128-4_4,
© Springer International Publishing Switzerland 2014

Fig. 4.1 Principle setup of an
actively Q-switched laser

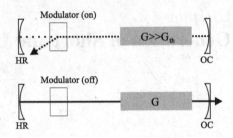

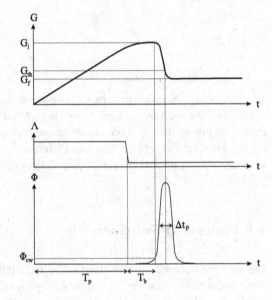

Fig. 4.2 Evolution of gain,
loss and photon density
during the Q-switch

Pumping at Low Q-Factor

During the pump phase of duration T_p the cavity losses are assumed to be high
enough to prevent lasing at all, i.e. $\langle \Phi \rangle \approx 0$. Thus, Eq. (2.60) may be written as

$$\frac{\partial \langle \Delta N \rangle}{\partial t} = R_p - \frac{\langle N \rangle + \langle \Delta N \rangle}{\tau}, \tag{4.1}$$

with the pump rate

$$R_p = 2 \frac{\lambda_p}{hc} I_p \frac{\eta_{abs}}{L}. \tag{4.2}$$

This can be easily solved under the assumption of a constant pump rate, resulting in
an inversion build-up according to

$$\langle \Delta N \rangle(t) = R_p \tau \left(1 - e^{-\frac{t}{\tau}}\right) - \langle N \rangle. \tag{4.3}$$

It is interesting to note here that this build-up is identical to the charging of a capacitor C, as shown in Fig. 4.3. Q-switching in this sense can thus be seen as slowly

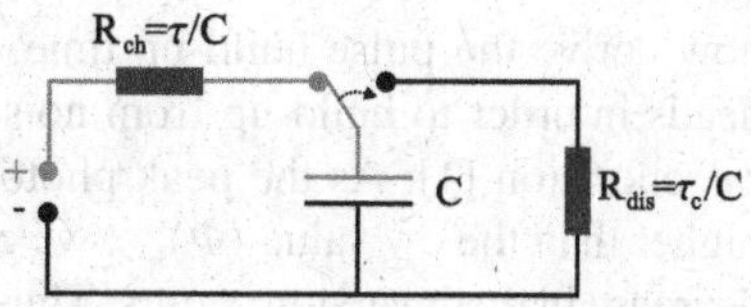

Fig. 4.3 Analogy between Q-switching and the charging of a capacitor

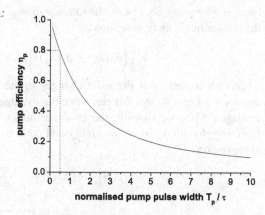

Fig. 4.4 Pump efficiency as a function of the pump pulse width

charging a capacitor over a high resistor $R_{ch} = \frac{\tau}{C}$ and quickly discharging it over a low resistor $R_{dis} = \frac{\tau_c}{C}$, which is connected to the much smaller cavity lifetime.

For a long time pumping, i.e. $t \to \infty$, the inversion will thus saturate and reach its upper limit

$$\langle \Delta N \rangle_{\infty} = R_p \tau - \langle N \rangle, \tag{4.4}$$

showing that long pumping phases will result in a low efficiency. To calculate the pump efficiency, we assume that the laser is pumped with a given pump energy E_p, which may be distributed over a variable pump time T_p in a square pulse. During this time $N_{p,max} = R_p T_p$ excitations will be created, which however, suffer from spontaneous decay. Therefore, at the end of the pump phase only

$$N_p = \langle \Delta N \rangle (T_p) + \langle N \rangle = \frac{N_{p,max}}{T_p} \tau \left(1 - e^{-\frac{T_p}{\tau}} \right) \tag{4.5}$$

excitations are still in the upper state. Thus the pump efficiency η_p can be derived as

$$\eta_p = \frac{\tau}{T_p} \left(1 - e^{-\frac{T_p}{\tau}} \right), \tag{4.6}$$

giving the amount of absorbed pump energy that is stored inside the laser medium excitation after the pump phase. As can be seen in Fig. 4.4, a pump pulse duration of $T_p < \frac{\tau}{2}$ should be used in order to get a pump efficiency $> 80\,\%$.

Pulse Build-Up at High Q-Factor

After the pump phase, the initial inversion $\langle \Delta N \rangle_i$ is present in the laser medium and the modulator is switched off, restoring the high Q-factor of the cavity. We will now derive the pulse build-up time, which is defined as the time the photon field needs in order to build-up from noise to a value comparable to the photon field in cw operation [1]. As the peak photon density in the Q-switch pulse will be much higher than the cw value $\langle \Phi \rangle_{cw}$, we can assume that for $\langle \Phi \rangle \le \langle \Phi \rangle_{cw}$ no significant decrease in the inversion occurs. Thus, the inversion is treated as constant during this time, and by using Eq. (2.63), the rate equation governing the temporal evolution of the photon field can be rewritten as

$$\frac{\partial \langle \Phi \rangle}{\partial t} = \frac{c}{2}\left[\sigma_a(\lambda_s) + \sigma_e(\lambda_s)\right]\left(\langle \Delta N \rangle_i - \langle \Delta N \rangle_{th}\right)\langle \Phi \rangle, \tag{4.7}$$

where, we assume that the axial changes in the population and the photon field are not too high, so that we can write the averaged products as the product of the averages. Also, we simplify the cross-sections of absorption and emission at the laser wavelength λ_s by $\sigma_a = \sigma_a(\lambda_s)$ and $\sigma_e = \sigma_e(\lambda_s)$ in the following. Using the abbreviations

$$\langle \Delta N \rangle_i' = \langle \Delta N \rangle_i - \frac{\sigma_a - \sigma_e}{\sigma_a + \sigma_e}\langle N \rangle \tag{4.8}$$

$$\langle \Delta N \rangle_{th}' = \langle \Delta N \rangle_{th} - \frac{\sigma_a - \sigma_e}{\sigma_a + \sigma_e}\langle N \rangle \tag{4.9}$$

$$r = \frac{\langle \Delta N \rangle_i'}{\langle \Delta N \rangle_{th}'} = \frac{g_i}{g_{th}}, \tag{4.10}$$

as well as Eq. (2.63) again we can simplify Eq. (4.7) to the form

$$\frac{\partial \langle \Phi \rangle}{\partial t} = \frac{1}{\tau_c}(r - 1)\langle \Phi \rangle, \tag{4.11}$$

with the solution

$$\langle \Phi \rangle(t) = \Phi_0 e^{(r-1)\frac{t}{\tau_c}}, \tag{4.12}$$

where, Φ_0 is the noise photon density caused by the vacuum fluctuations. The cavity field, therefore, will start growing exponentially from the vacuum fluctuations with the time constant $\frac{\tau_c}{r-1}$ until it depletes the inversion significantly. The pump parameter r can also be expressed as the ratio between the initial logarithmic gain g_i and the logarithmic threshold gain g_{th} using

$$g_i = (\sigma_a + \sigma_e)\langle \Delta N \rangle_i - (\sigma_a - \sigma_e)\langle N \rangle \tag{4.13}$$

$$g_{th} = (\sigma_a + \sigma_e)\langle \Delta N \rangle_{th} - (\sigma_a - \sigma_e)\langle N \rangle. \tag{4.14}$$

As long as the depletion of the ground-state N_1 can be neglected during pumping, e.g. in high-repetition rate operation as discussed later on, the logarithmic gain is proportional to the pump power, resulting in

$$r = \frac{g_i}{g_{th}} \approx \frac{P_p}{P_{th}}. \tag{4.15}$$

Therefore, the pump parameter r is often identified with the "times-above-threshold" operation point of the laser given by $r - 1$.

Fig. 4.5 Pulse build-up time as a function of the pump parameter r

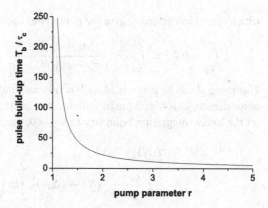

Defining the cavity build-up time T_b by $\langle \Phi \rangle (T_b) = \langle \Phi \rangle_{cw}$ results in

$$T_b = \frac{\tau_c}{r-1} \ln \frac{\langle \Phi \rangle_{cw}}{\Phi_0}. \tag{4.16}$$

In most laser systems, the ratio between the cw photon density and the noise is of the order of $10^8 - 10^{12}$, giving

$$T_b \approx (22.5 \pm 5) \frac{\tau_c}{r-1}. \tag{4.17}$$

As shown in Fig. 4.5, the pulse build-up time quickly decreases with increasing pump power, shifting towards the time when the modulator opens. In order not to loose efficiency, additional losses from the modulator must be avoided. Therefore, the modulator has to be chosen so that the switching between the low-Q and the high-Q state of the cavity occurs much faster than the build-up time of the laser pulse.

Pulse Peak Power and Pulse Width

To derive the pulse width of the Q-switch pulse, we can assume that during the pulse build-up and the pulse extraction time, we can neglect further spontaneous decay of the upper level as well as pumping, which results in the rate equations

$$\frac{\partial \langle \Delta N \rangle}{\partial t} = c \big[(\sigma_a - \sigma_e) \langle N \rangle - (\sigma_a + \sigma_e) \langle \Delta N \rangle \big] \langle \Phi \rangle \tag{4.18}$$

$$\frac{\partial \langle \Phi \rangle}{\partial t} = \frac{c}{2} (\sigma_a + \sigma_e) \big(\langle \Delta N \rangle - \langle \Delta N \rangle_{th} \big) \langle \Phi \rangle. \tag{4.19}$$

Dividing Eq. (4.19) by Eq. (4.18) yields the evolution of the photon field with inversion as

$$\frac{\partial \langle \Phi \rangle}{\partial \langle \Delta N \rangle} = \frac{1}{2} \frac{(\sigma_a + \sigma_e)(\langle \Delta N \rangle - \langle \Delta N \rangle_{th})}{(\sigma_a - \sigma_e) \langle N \rangle - (\sigma_a + \sigma_e) \langle \Delta N \rangle}, \tag{4.20}$$

which can be integrated to give the photon field as a function of the inversion density,

$$2\int_{\Phi_0}^{\langle\Phi\rangle} d\langle\Phi\rangle = \int_{\langle\Delta N\rangle_i}^{\langle\Delta N\rangle} \frac{[\sigma_a+\sigma_e](\langle\Delta N\rangle-\langle\Delta N\rangle_{th})}{([\sigma_a-\sigma_e]\langle N\rangle-[\sigma_a+\sigma_e]\langle\Delta N\rangle)} d\langle\Delta N\rangle. \quad (4.21)$$

This integral can be performed analytically and under the assumption that the photon noise density is low compared with the one occurring during the pulse, i.e. we can set the lower integration boundary to $\Phi_0 \approx 0$, this results in

$$2\langle\Phi\rangle \approx \langle\Delta N\rangle_i - \langle\Delta N\rangle$$
$$+ \left[\frac{\sigma_a-\sigma_e}{\sigma_a+\sigma_e}\langle N\rangle - \langle\Delta N\rangle_{th}\right] \ln\left(\frac{\langle\Delta N\rangle_i - \frac{\sigma_a-\sigma_e}{\sigma_a+\sigma_e}\langle N\rangle}{\langle\Delta N\rangle - \frac{\sigma_a-\sigma_e}{\sigma_a+\sigma_e}\langle N\rangle}\right). \quad (4.22)$$

After the pulse is emitted the photon density will decrease to zero again and a residual (final) inversion $\langle\Delta N\rangle_f$ is left inside the medium given by the relation

$$\langle\Delta N\rangle_f - \langle\Delta N\rangle_i$$
$$= \left[\frac{\sigma_a-\sigma_e}{\sigma_a+\sigma_e}\langle N\rangle - \langle\Delta N\rangle_{th}\right] \ln\left(\frac{\langle\Delta N\rangle_i - \frac{\sigma_a-\sigma_e}{\sigma_a+\sigma_e}\langle N\rangle}{\langle\Delta N\rangle_f - \frac{\sigma_a-\sigma_e}{\sigma_a+\sigma_e}\langle N\rangle}\right). \quad (4.23)$$

This is the main equation describing the Q-switch process. Using the abbreviations in Eqs. (4.8)–(4.10) and accordingly

$$\langle\Delta N\rangle_f' = \langle\Delta N\rangle_f - \frac{\sigma_a-\sigma_e}{\sigma_a+\sigma_e}\langle N\rangle \quad (4.24)$$

the fundamental Q-switch equation can be rewritten in the simple form

$$\frac{\langle\Delta N\rangle_f'}{\langle\Delta N\rangle_i'} = 1 - \frac{1}{r}\ln\frac{\langle\Delta N\rangle_i'}{\langle\Delta N\rangle_f'}, \quad (4.25)$$

showing that the whole Q-switch pulse evolution only depends on the initial inversion $\langle\Delta N\rangle_i'$ and the cavity parameters included in $\langle\Delta N\rangle_{th}'$.

To derive the pulse peak power, we first have to find the time of the pulse peak itself. As already shown in Fig. 4.2, the peak is reached when no further net amplification is possible, i.e. it will occur exactly when the gain, and thus the inversion, crosses the threshold values. Using Eq. (4.22) thus gives the peak photon density inside the cavity as

$$\langle\hat{\Phi}\rangle = \frac{r-1-\ln r}{2}\langle\Delta N\rangle_{th}'. \quad (4.26)$$

Therefore, it only depends on the cavity parameters and r. As these photons will leave the cavity with the cavity photon lifetime τ_c, the peak power of the Q-switched pulse can be directly given by

$$\hat{P} = \frac{h\nu}{\tau_c}\langle\hat{\Phi}\rangle V = \frac{r-1-\ln r}{2}\langle\Delta N\rangle_{th}'\frac{h\nu}{\tau_c}V. \quad (4.27)$$

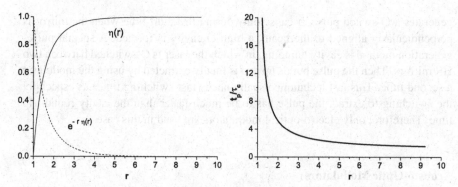

Fig. 4.6 Extraction efficiency and relative pulse width of a Q-switch pulse as a function of the pump parameter r

Additionally, we define the energy extraction efficiency η_e by the fraction of extracted inversion as

$$\eta_e = 1 - \frac{\langle \Delta N \rangle'_f}{\langle \Delta N \rangle'_i}. \tag{4.28}$$

Using Eq. (4.25) the energy extraction efficiency $\eta_e(r)$ can be calculated independently from the actual laser parameters by the transcendental equation

$$r = -\frac{\ln[1 - \eta_e(r)]}{\eta_e(r)}. \tag{4.29}$$

Hence, we can approximate the pulse width t_p of the Q-switch pulse as the ratio between the extracted energy $E_s = \frac{1}{2}h\nu V(\langle \Delta N \rangle'_i - \langle \Delta N \rangle'_f)$ and the pulse peak power $\hat{P}$ by

$$t_p \approx \frac{E_s}{\hat{P}} = \frac{r\eta_e(r)}{r - 1 - \ln r}\tau_c. \tag{4.30}$$

The factor $\frac{1}{2}$ in the energy takes into account that in the ΔN each excitation is counted twice.

As can be seen in Fig. 4.6 the extraction efficiency quickly approaches unity for $r > 4$, whilst the pulse width asymptotically decreases towards the cavity lifetime. This shows that short pulses on the order on several ns to 1 μs are possible with Q-switched lasers, depending on the cavity lengths and lifetimes.

4.1.2 Experimental Realization

Q-switching is most often achieved through use of two main techniques, in which either an acousto-optic modulator (AOM) or an electro-optic modulator (EOM) is used to modify the cavity losses. The initial technique was to rotate the HR mirror of the cavity around an axis perpendicular to the beam propagation axis. This generates a Q-switch pulse, because only during the short time when the mirror is perpendicularly aligned to the beam, a high-Q cavity is formed. A special pulse-generation method is cavity dumping, in which the laser is Q-switched between two HR mirrors. Then the pulse builds up and is finally extracted by using the modulator a second time. This last technique usually needs fast switching times as especially the switching to extract the pulse has to be much faster than the cavity round-trip time. Therefore, only electro-optical modulators are used in this case.

Acousto-Optic Modulators

The usual setup of an acousto-optically Q-switched laser is shown in Fig. 4.7. The modulator consists of a transparent material, e.g. silica glass (SiO_2) or tellurium dioxide (TeO_2), to which an ultrasonic transducer is bonded to create a sound wave inside the bulk modulator material. Owing to the photo-elastic effect, this sound wave generates an index of refraction distribution inside the modulator material, which behaves as an optical phase grating that causes a part of the incident power to be diffracted out of the cavity, thus creating losses. By switching off the radio-frequency (rf) power to the transducer, the glass block returns to its homogeneous index state and the high Q-factor of the resonator is restored [3].

Depending on the length L_m of the modulator material, the wavelengths of the optical wave and the sound wave, two diffraction regimes are observed, which are the Raman-Nath regime and the Bragg regime.

In **Raman-Nath scattering** the interaction length L_m is short or the sound wave-length λ_a is large, thus $\lambda_s L_m \ll \lambda_a^2$. In this case, the incident light is scattered into many diffraction orders, with a maximum of diffracted power occurring when the sound wave interacts perpendicularly with the light wave, as shown in Fig. 4.8. The amplitude of the phase grating is given by

$$\Delta\phi = 2\pi \Delta n \frac{L_m}{\lambda_s} = \pi \sqrt{\frac{2L_m}{\lambda_s^2} M_2 \frac{P_a}{b}}, \qquad (4.31)$$

with b being the width of the sound wave, P_a the acoustic wave power and M_2 the so-called figure of merit of the acousto-optic material. It can be calculated from the refractive index n, the photoelastic coefficient in the chosen geometry p, the density of the acousto-optic material ρ and the velocity of sound v_a as

$$M_2 = \frac{n^6 p^2}{\rho v_a^3}. \qquad (4.32)$$

Finally, the intensity scattered into the nth order is given by

$$I_n = \hat{I}_0 J_n^2(\Delta\phi), \qquad (4.33)$$

where, $J_n(x)$ is the Bessel function of nth order and $\hat{I}_0$ the incident laser intensity.

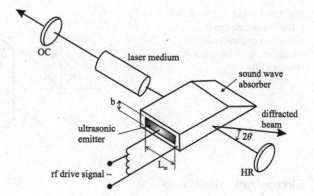

Fig. 4.7 Setup of an actively Q-switched laser using an acousto-optic modulator [3]

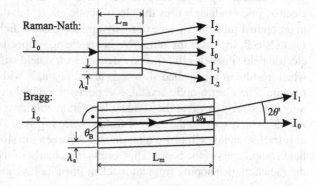

Fig. 4.8 The two operation regimes of an acousto-optic modulator: The Raman-Nath regime and the Bragg regime [3]

In **Bragg scattering**, described by $\lambda_s L_m \gg \lambda_a^2$, a zero-order and first-order diffraction beam become predominant under the Bragg condition [3], in this case the sound wave and the light wave interact at the **Bragg angle** θ_B, given by

$$\sin\theta_B = \frac{\lambda_s}{2n\lambda_a}. \tag{4.34}$$

The internal deflection angle is given by $2\theta_B$ and by taking into account the refraction on the output side of the modulator, one finds an external diffraction angle of

$$\theta' = 2n\theta_B \approx \frac{\lambda_s}{\lambda_a}. \tag{4.35}$$

The intensity of the scattered beam is then given by

$$I_1 = \hat{I}_0 \sin^2 \frac{\Delta\phi}{2}, \tag{4.36}$$

and the intensity of the transmitted beam I_0 is reduced by this amount compared with the off-state of the modulator.

Fig. 4.9 Layout of a Pockels cell as an electro-optic modulator and the induced change in the refractive index ellipsoid [3]

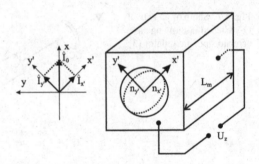

Electro-Optic Modulators

Whilst acousto-optic modulators may also be used with unpolarized light, an electro-optic modulator uses the electro-optic effect, i.e. the birefringence induced in an optical medium by an externally applied electric field. This is achieved in a **Pockels cell**, in which the refractive index change depends linearly on the applied electric field (**Pockels effect**). The external electric field will induce a birefringence, which results in a so-called slow-axis and a fast-axis with different indices of refraction. The electro-optic crystal, e.g. potassium dihydrogen phosphate (KDP), is oriented in such way that the incident laser light will have its polarized aligned under 45° with respect to the slow or fast axis, see Fig. 4.9. Then, the induced change in refractive index will cause a phase shift between the slow- and fast-axis electric field components of the beam. This results in a change of the state of polarization of the radiation, developing from an incident linear polarization to an elliptical polarization and a circular polarization during its propagation along the cell axis.

For a given cell length L_c two specific voltages exist for which the output polarization corresponds to a circular polarization or a linear polarization rotated by 90° with respect to the incident polarization orientation. These voltages are called quarter-wave $U_{\frac{\lambda}{4}}$ and half-wave voltage $U_{\frac{\lambda}{2}}$, respectively,

$$U_{\frac{\lambda}{4}} = \frac{\lambda_s}{4n_0^3 r_{63}}, \tag{4.37}$$

$$U_{\frac{\lambda}{2}} = \frac{\lambda_s}{2n_0^3 r_{63}}, \tag{4.38}$$

as the cell acts like a quarter- or half-wave plate in this case. In this formulae, wherein n_0 is the ordinary index of refraction, λ_s is the laser wavelength and r_{63} is the electro-optic coefficient. Combining such a Pockels cell with an intracavity polarizer now allows efficient and fast switching of the internal beam as the electro-optic effect has a response time much smaller than the cavity time constants. The switching time only depends on the high-voltage power supply and its ability to charge the Pockels cell, which is electrically charged just like a capacitor.

The quarter-wave setup only needs one intra-cavity polarizer, since the beam passes the Pockels cell twice resulting in a total polarization rotation of 90° as shown in Fig. 4.10. In the half-wave setup a second polarizer is needed to couple out the

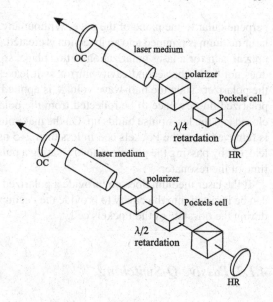

Fig. 4.10 Setup of an actively Q-switched laser using an electro-optic modulator [3]

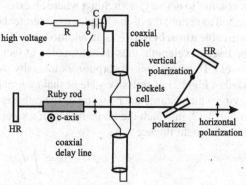

Fig. 4.11 Setup of a cavity-dumped ruby laser [3]

all the incident radiation when the voltage is applied. By switching off the voltage in both cases the electro-optic crystal will return into its non-birefringent state and the cavity is restored, causing the Q-switch pulse to build up.

Cavity Dumping

In cavity dumping the half-wave setup of a Pockels cell is used and the laser is Q-switched with nearly 100 % reflectivity cavity mirrors in order to obtain very short Q-switch pulses. At the peak of the Q-switched pulse, the Pockels cell is used to switch the closed cavity rapidly to its output port, provided by an intracavity polarizer. Thus, the width of the Q-switched pulse is only a function of the cavity length and its round-trip time, and not of the spectroscopic parameters of the laser medium. In the example in Fig. 4.11 the ruby laser rod is oriented so that the c-axis is perpendicular to the plane of the page. Without any voltage on the Pockels cell, the laser medium is pumped and an inversion is created. The crystal only provides a high optical gain for a laser polarization in this plane, so that the generated fluorescence does not "see" the second cavity mirror as it leaves the cavity by passing through the polarizer. Then the half-wave voltage is applied to the Pockels cell, causing the polarized fluorescence to be reflected from the polarizer. Hence, the laser cavity is closed and the laser pulse builds up. On the maximum laser pulse power, the voltage is removed from the Pockels cell in less than 2–5 ns and the cavity photons will all leak out by passing the polarizer, thus creating a pulse with a width of the round-trip time of the resonator.

If the laser medium does not provide a polarized output itself, a second polarizer can be inserted into the cavity to provide the decoupling of the second cavity mirror during the off-state of the Pockels cell.

4.1.3 Passive Q-Switching

In contrast to active Q-switching, where an external signal is applied to open the cavity and to restore the high Q-factor to generate the pulse, passive Q-switching uses a **saturable absorber**. This is an additional medium inside the cavity that absorbs on the laser wavelength, thus decreasing the Q-factor (or increasing the internal cavity losses). However, this absorption is intensity dependent and quickly saturates towards a highly transmissive state of that material, restoring the high Q-factor of the cavity, which causes the build-up of the pulse intensity. This switching can be seen in Fig. 4.12, in which the transmission of a saturable medium is shown with respect to the incident fluence

$$J = \int I_s dt \qquad (4.39)$$

on the absorption line. By analogy with the Frantz-Nodvik model [2], this transmission can be calculated by

$$T(J) = \frac{J_{sat}}{J} \ln\left[1 + \left(e^{\frac{J}{J_{sat}}} - 1\right)T_0\right], \qquad (4.40)$$

where, T_0 is the initial, i.e. unpumped, transmission of the saturable medium and J_{sat} is the saturation fluence, given by

$$J_{sat} = \frac{hc}{\lambda_s[\sigma_a(\lambda_s) + \sigma_e(\lambda_s)]} = \tau^* I_{sat}^s, \qquad (4.41)$$

with τ^* being the excitation lifetime of the saturable absorber and

$$I_{sat}^s = \frac{hc}{\lambda_s[\sigma_a(\lambda_s) + \sigma_e(\lambda_s)]\tau^*} \qquad (4.42)$$

being the saturation intensity of the absorber on the laser line. It should not be confused with the pump saturation intensity in Eq. (2.71), in which the pump wavelength λ_p occurs.

Fig. 4.12 Transmission of a saturable medium as a function of the incident fluence at an initial transmission of $T_0 = 0.6$

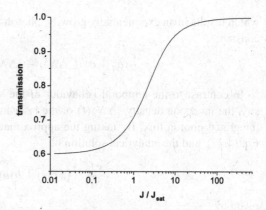

Owing to the additional saturable absorber inside the cavity, a new rate equation has to be added to describe this system. The passive Q-switch on the time scale of the pulse generation, i.e. when pumping and spontaneous decay can be neglected, is therefore given by the coupled equations

$$\frac{\partial \langle \Delta N \rangle}{\partial t} = c\left[(\sigma_a - \sigma_e)\langle N \rangle - (\sigma_a + \sigma_e)\langle \Delta N \rangle\right]\langle \Phi \rangle \tag{4.43}$$

$$\frac{\partial \Delta N^*}{\partial t} = c\left[(\sigma_a^* - \sigma_e^*)N^* - (\sigma_a^* + \sigma_e^*)\Delta N^*\right]\langle \Phi \rangle - \frac{\Delta N^* + N^*}{\tau^*} \tag{4.44}$$

$$\frac{\partial \langle \Phi \rangle}{\partial t} = \frac{c}{2}(\sigma_a + \sigma_e)(\langle \Delta N \rangle - \langle \Delta N \rangle_{th})\langle \Phi \rangle$$

$$+ \frac{c}{2}\left[(\sigma_a^* + \sigma_e^*)\Delta N^* - (\sigma_a^* - \sigma_e^*)N^*\right]\langle \Phi \rangle, \tag{4.45}$$

where, ΔN^* and N^* are the inversion density and total absorber density of the saturable absorber, τ^* its excitation lifetime and $\sigma_a^* = \sigma_a^*(\lambda_s)$ and $\sigma_e^* = \sigma_e^*(\lambda_s)$ the saturable absorber cross-sections of absorption and emission at the laser wavelength λ_s.

A saturable absorber usually has a very low excitation lifetime $\tau^* <$ ns, much lower than the Q-switch pulse widths created. It often uses dyes or semiconductor materials. Thus, the inversion density of the saturable absorber ΔN^* in Eq. (4.44) will nearly instantaneously react on the photon density $\langle \Phi \rangle$. Therefore, we can approximately solve this rate equation as being in the steady state compared to all other processes during the Q-switch. This results in

$$\Delta N^* = \frac{c\tau^*(\sigma_a^* - \sigma_e^*)\langle \Phi \rangle - 1}{c\tau^*(\sigma_a^* + \sigma_e^*)\langle \Phi \rangle + 1}N^*. \tag{4.46}$$

At the beginning of the Q-switch process, it can be assumed that the laser medium has its initial inversion density $\langle \Delta N \rangle_i$ and that the saturable absorber is still unexcited, i.e. $\Delta N^* \approx -N^*$. Therefore, Eq. (4.45) gives

$$\frac{\partial \langle \Phi \rangle}{\partial t} = \frac{c}{2}(\sigma_a + \sigma_e)(\langle \Delta N \rangle_i - \langle \Delta N \rangle_{th})\langle \Phi \rangle - c\sigma_a^* N^* \langle \Phi \rangle, \tag{4.47}$$

which results in an exponentially growing photon field $\langle\Phi\rangle(t) = \Phi_0 e^{\gamma_0 t}$ with a time constant

$$\gamma_0 = \frac{c}{2}(\sigma_a + \sigma_e)(\langle\Delta N\rangle_i - \langle\Delta N\rangle_{th}) - c\sigma_a^* N^*. \tag{4.48}$$

In contrast to the temporal behaviour of the saturable absorber inversion density, the inversion density $\langle\Delta N\rangle(t)$ of the laser medium will be determined by the integrated photon flux. By taking the approximate exponential growth solution of Eq. (4.47), and the analytical solution of

$$\frac{\partial f}{\partial t} = \big(af(t) + b\big)u(t), \tag{4.49}$$

given by

$$f(t) = e^{a\int_0^t u(t')dt'}\left(f(0) + b\int_0^t e^{-a\int_0^{t'} u(t'')dt''}u(t')dt'\right), \tag{4.50}$$

Eq. (4.43) can be analytically solved, giving

$$\langle\Delta N\rangle = e^{-\frac{c(\sigma_a+\sigma_e)}{\gamma_0}\langle\Phi\rangle(t)}\left(\langle\Delta N\rangle_i + c(\sigma_a - \sigma_e)\langle N\rangle\int_0^t e^{\frac{c(\sigma_a+\sigma_e)}{\gamma_0}\langle\Phi\rangle(t')}\langle\Phi\rangle(t')dt'\right). \tag{4.51}$$

Inserting these results into Eq. (4.45), the exponential time constant of the photon field can be described by

$$\frac{1}{\langle\Phi\rangle}\frac{\partial\langle\Phi\rangle}{\partial t} = \gamma_0 + \left(c^2\sigma_a^*(\sigma_a^* + \sigma_e^*)\tau^* N^* - \frac{c^2(\sigma_a+\sigma_e)^2}{2\gamma_0}\langle\Delta N\rangle_i\right)\langle\Phi\rangle + \cdots, \tag{4.52}$$

wherein a series development in the power of $\langle\Phi\rangle$ was used.

If the coefficient of the linear term in $\langle\Phi\rangle$ has a negative sign, the exponential time constant will decrease with increasing photon flux, which means that the gain provided by the laser medium saturates before the absober can saturate. This will thus not result in a Q-switch pulse. However, when the sign of this linear term is positive, the exponential time constant will increase with increasing photon flux as the saturable absorber bleaches much faster than the gain of the laser medium is reduced owing to amplification. Then, a Q-switch pulse is emitted as shown in Fig. 4.13. Passive Q-switching thus depends on two thresholds: a first threshold that needs to be passed by pumping strongly enough that the generated gain exceeds the unsaturated losses of the cavity including the saturable absorber, and a second threshold that is given by passing the point after which the photon flux grows faster than exponentially. If we denote the single-pass gain before saturation occurs with G_0, the logarithmic round-trip gain results in $g_0 = 2\ln G_0$. Hence, the exponential time constant γ_0 may be approximated by

$$\gamma_0 \approx \frac{g_0}{\Delta t_{RT}}, \tag{4.53}$$

Fig. 4.13 Evolution of gain, loss and photon density during the passive Q-switch

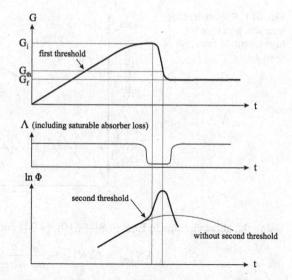

neglecting the cavity photon lifetime, see Eq. (2.62). Therein, Δt_{RT} is the cavity round-trip time. Then, the **second threshold** can be expressed by

$$N^* > \frac{(\sigma_a + \sigma_e)^2}{\sigma_a^*(\sigma_a^* + \sigma_e^*)} \frac{\Delta t_{RT}}{\tau^*} \frac{\langle \Delta N \rangle_i}{2g_0},$$ (4.54)

stating the minimum absorber density necessary to pass the second threshold.

4.1.4 Scaling Laws of Repetitive Q-Switching

In this section we will investigate repetitive Q-switching, i.e. a periodic opening and closing of the cavity by the modulator at a repetition rate ν_{Rep} and with an opening time t_G, called a gate. Of-course, the gate t_G has to be at least as long as the pulse build-up time. As a result from the finite pulse build-up time an upper limit will exist for the repetition rate, given by the fact that during the corresponding repetition period $T_{Rep} = \frac{1}{\nu_{Rep}}$ enough inversion, and thus, gain has to build-up so that the pulse will be created within the gate duration, i.e. during the high-Q state of the cavity.

In repetitive Q-switching under equilibrium conditions, i.e. when all pulses show equal pulse energy, it follows from the dependence of the Q-switch pulse evolution in Eq. (4.25) that the initial inversion before each pulse emission has to be equal. As the initial inversion of the nth pulse is coupled to the final inversion of the $n - 1^{th}$ pulse, by the pumping between the two pulses, we can conclude from self-consistency, using Eq. (4.1), that

$$\langle \Delta N \rangle_i = \left(\langle \Delta N \rangle_f - R_p \tau + \langle N \rangle \right) e^{-\frac{T_{Rep}}{\tau}} + R_p \tau - \langle N \rangle$$

$$= \langle \Delta N \rangle_\infty - \left(\langle \Delta N \rangle_\infty - \langle \Delta N \rangle_f \right) e^{-\frac{1}{\nu_{Rep}\tau}}.$$ (4.55)

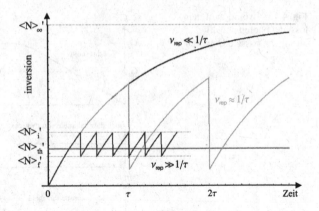

Fig. 4.14 Evolution of the inversion with time for high-repetition-rate Q-switching

Using the abbreviations in Eqs. (4.8)–(4.10), (4.24) and accordingly

$$\langle \Delta N \rangle'_\infty = \langle \Delta N \rangle_\infty - \frac{\sigma_a - \sigma_e}{\sigma_a + \sigma_e} \langle N \rangle, \qquad (4.56)$$

Eq. (4.55) can be rewritten to

$$\langle \Delta N \rangle'_i = \langle \Delta N \rangle'_\infty - \left(\langle \Delta N \rangle'_\infty - \langle \Delta N \rangle'_f \right) e^{-\frac{1}{\nu_{Rep}\tau}}. \qquad (4.57)$$

The other two equations necessary to derive the scaling laws are rewritten forms of Eqs. (4.25), (4.30) and are given by

$$\langle \Delta N \rangle'_i - \langle \Delta N \rangle'_f = \langle \Delta N \rangle'_{th} \ln \frac{\langle \Delta N \rangle'_i}{\langle \Delta N \rangle'_f}, \qquad (4.58)$$

$$\Delta t_p = \frac{\langle \Delta N \rangle'_i - \langle \Delta N \rangle'_f}{\langle \Delta N \rangle'_i - \langle \Delta N \rangle'_{th}(1 + \ln \frac{\langle \Delta N \rangle'_i}{\langle \Delta N \rangle'_{th}})} \tau_c. \qquad (4.59)$$

In the case of low repetition rates, i.e. $\nu_{Rep} \ll \frac{1}{\tau}$, Eq. (4.57) yields $\langle \Delta N \rangle'_i \approx \langle \Delta N \rangle'_\infty$ and thus, using Eq. (4.58), that $\langle \Delta N \rangle'_f \approx$ constant. Therefore, also the pulse width Δt_p, pulse peak power $\hat{P}$ and pulse energy E_s are constant and the average power, given by

$$\langle P_s \rangle = \frac{1}{2} h \nu V \left(\langle \Delta N \rangle'_i - \langle \Delta N \rangle'_f \right) \nu_{Rep}, \qquad (4.60)$$

scales with the repetition rate.

For high repetition rates, i.e. $\nu_{Rep} \gg \frac{1}{\tau}$, this calculation is a little bit more complex. In this case, we can assume $\langle \Delta N \rangle'_i \approx \langle \Delta N \rangle'_f$, as shown in Fig. 4.14, and we can thus develop the logarithm in Eq. (4.58) to third order,

$$\ln x \simeq -\frac{(x-1)^2}{2} + x - 1, \qquad (4.61)$$

resulting in

$$\frac{\langle \Delta N \rangle'_f}{\langle \Delta N \rangle'_{th}} \simeq 2 - \frac{\langle \Delta N \rangle'_i}{\langle \Delta N \rangle'_f}. \qquad (4.62)$$

Table 4.1 Scaling laws of the repetitively Q-switched laser

Repetition rate	Average power	Pulse width	Peak power	Pulse energy
$\nu_{Rep} \ll \frac{1}{\tau}$	$\langle P_s \rangle \propto \nu_{Rep}$	$\Delta t \sim \text{const}$	$\hat{P} \sim \text{const}$	$E_s \sim \text{const}$
$\nu_{Rep} \gg \frac{1}{\tau}$	$\langle P_s \rangle \sim \text{const}$	$\Delta t \propto \nu_{Rep}$	$\hat{P} \propto \frac{1}{\nu_{Rep}^2}$	$E_s \propto \frac{1}{\nu_{Rep}}$

As we can also assume $\langle \Delta N \rangle_i' \approx \langle \Delta N \rangle_{th}'$, we can use the same third-order development in Eq. (4.59) and insert the result of Eq. (4.62), giving

$$\Delta t_p \simeq \tau_c \frac{\langle \Delta N \rangle_i' - \langle \Delta N \rangle_f'}{\frac{\langle \Delta N \rangle_{th}'}{2}(1 - \frac{\langle \Delta N \rangle_i'}{\langle \Delta N \rangle_f'})^2}. \tag{4.63}$$

Using the equivalent of Eq. (4.58),

$$\frac{\langle \Delta N \rangle_i'}{\langle \Delta N \rangle_f'} = e^{\frac{\langle \Delta N \rangle_i' - \langle \Delta N \rangle_f'}{\langle \Delta N \rangle_{th}'}} \simeq 1 + \frac{\langle \Delta N \rangle_i' - \langle \Delta N \rangle_f'}{\langle \Delta N \rangle_{th}'} \tag{4.64}$$

for $\frac{\langle \Delta N \rangle_i' - \langle \Delta N \rangle_f'}{\langle \Delta N \rangle_{th}'} \ll 1$, we can deduce

$$\Delta t_p \simeq \frac{2\tau_c \langle \Delta N \rangle_{th}'}{\langle \Delta N \rangle_i' - \langle \Delta N \rangle_f'}. \tag{4.65}$$

As $\nu_{Rep} \gg \frac{1}{\tau}$, it follows from Eq. (4.57) that

$$\langle \Delta N \rangle_i' - \langle \Delta N \rangle_f' \simeq \frac{\langle \Delta N \rangle_\infty' - \langle \Delta N \rangle_f'}{\tau \nu_{Rep}}, \tag{4.66}$$

and from $\langle \Delta N \rangle_f' \ll \langle \Delta N \rangle_\infty'$ we finally obtain

$$\Delta t_p \propto \frac{\nu_{Rep}}{\langle \Delta N \rangle_\infty'}. \tag{4.67}$$

Thus, the pulse width will increase linearly with repetition rate for a constant pump power, and it will decrease with increasing pump power, i.e. with increasing $\langle \Delta N \rangle_\infty'$. Using Eq. (4.60) and the relation

$$\hat{P} = \frac{\langle P_s \rangle}{\Delta t_p \nu_{Rep}} \tag{4.68}$$

we obtain the other scaling laws shown in Table 4.1.

As for high repetition rates, the average output power is constant; this regime of operation is also often called **quasi-continuous operation**. Resulting from the linear increase in pulse width, as well as the fact that with increasing repetition rate, the pulse energy is distributed over an increasing number of pulses, the peak power will strongly decrease with the inverse square of the repetition rate. For low repetition rate operation, the continuous pumping will saturate the inversion and the initial inversion becomes pump duration, i.e. repetition period, independent. Therefore, every pulse has the maximum pulse energy given by the completely inverted

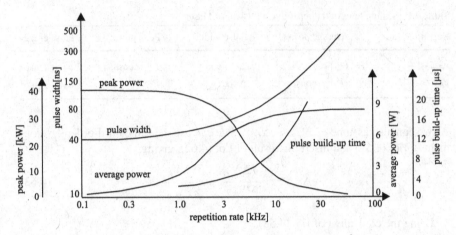

Fig. 4.15 Evolution of the laser output parameters with repetition rate for a continuously-pumped Q-switched laser [3]

population in the laser medium and the average output power simply increases with repetition rate. However, it has to be noted here that this case is usually difficult to achieve. In most lasers the fully inverted laser medium corresponds to such a high pulse energy that the optical damage threshold of the coatings on some intracavity components, such as the mirrors or the laser medium itself will be exceeded, resulting in the destruction of this component. The transition between the two repetition frequency regimes is non-linear and makes a numerical solution of the rate equations necessary. In summary, the dependence of the output parameters of a continuously-pumped Q-switched Nd^{3+}:YVO_4 laser is shown in Fig. 4.15.

4.2 Basics of Mode Locking and Ultra-Short Pulses

As we investigated in the previous chapter, short laser pulses on the order of the cavity lifetime τ_c, i.e. with a duration of several ns to μs, can be created with the Q-switch technique. In a careful design the laser, these pulses may correspond to a single longitudinal mode. If much shorter pulses are necessary, the longitudinal mode structure of the laser needs to be exploited, as pulse width and laser spectrum are coupled by an uncertainty-like relation. To investigate this, we consider a Gaussian laser pulse with an electric field amplitude

$$E(t) = E_0 e^{-\xi t^2} e^{i\omega_0 t} \tag{4.69}$$

with a Gaussian parameter

$$\xi = a - ib. \tag{4.70}$$

Thus, the laser pulse intensity $I(t) \propto |E(t)|^2$ will be

$$I(t) = I_0 e^{-4\ln 2(\frac{t}{\tau_p})^2}, \tag{4.71}$$

Fig. 4.16 Electric field of a chirped Gaussian pulse

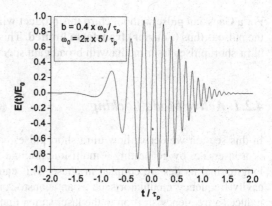

where, the pulse width τ_p is given by

$$\tau_p = \sqrt{\frac{2\ln 2}{a}}. \tag{4.72}$$

We can interpret $a = \Re(\xi)$ as being connected to the pulse width, while $b = \Im(\xi)$ is connected to the **chirp** of the pulse, i.e. the time-dependent frequency shift during the pulse. This can be seen directly from Eq. (4.69), which results in a total pulse phase given by

$$\phi = \omega_0 t + bt^2, \tag{4.73}$$

so that the actual laser frequency is given by

$$\omega = \frac{\partial \phi}{\partial t} = \omega_0 + 2bt. \tag{4.74}$$

Therefore, b describes a linear chirp, i.e. a linearly increasing laser frequency during the pulse, as shown in Fig. 4.16.

In order to derive the relation between laser pulse width and the spectral output, the frequency spectrum of the electric field is calculated by its Fourier transform,

$$\tilde{E}(\omega) = \tilde{E}_0 e^{-\frac{(\omega-\omega_0)^2}{4\xi}} = \tilde{E}_0 e^{-\frac{1}{4}(\frac{a}{a^2+b^2}+i\frac{b}{a^2+b^2})(\omega-\omega_0)^2}. \tag{4.75}$$

Therefore, the spectral intensity distribution $\tilde{I}(\omega) \propto |\tilde{E}(\omega)|^2$ is given by

$$\tilde{I}(\omega) = \tilde{I}_0 e^{-\frac{1}{2}\frac{a}{a^2+b^2}(\omega-\omega_0)^2} = \tilde{I}_0 e^{-4\ln 2(\frac{\omega-\omega_0}{2\pi\Delta\nu_p})^2}. \tag{4.76}$$

Thus, the pulse bandwidth results in

$$\Delta\nu_p = \frac{\sqrt{2\ln 2}}{\pi} \sqrt{a\left(1+\left(\frac{b}{a}\right)^2\right)}, \tag{4.77}$$

and the time-bandwidth product is given by

$$\tau_p \Delta\nu_p = \frac{2\ln 2}{\pi} \sqrt{1+\left(\frac{b}{a}\right)^2} \approx 0.44 \sqrt{1+\left(\frac{b}{a}\right)^2}. \tag{4.78}$$

For a Gaussian pulse without chirp this product will be given by $\tau_p \Delta f_p = 0.44$ and the pulse is thus **(Fourier) transform limited**. This shows that for the generation of ultra-short pulses laser media with broad gain spectra are necessary.

4.2.1 Active Mode Locking

In this section we derive how ultra-short pulses can be obtained by mode-locking of a laser, i.e. by generating a multi-longitudinal mode emission in which all the longitudinal modes are coupled in phase. This can be obtained by use of an intra-cavity frequency modulator such as an acousto- or electro-optic modulator, which induces a frequency shift on to the laser signal that corresponds exactly to the free-spectral range, and thus, the mode spacing of the cavity. Let us assume that the laser starts oscillating on the strongest line first, corresponding to a longitudinal mode index q_0. Then, after passing through the modulator, a fraction of the laser power will be shifted towards the modes $q_0 \pm 1$, that can be seen as sidebands to the main mode and which are also amplified, as the gain spectrum is assumed to be broad. As this shifted fraction usually has a much higher intensity than the spontaneous emission at that wavelength, the laser medium will predominantly amplify these shifted photons, which have a unique phase relation to the central mode q_0 with a phase difference ϕ. The amplified sidebands get shifted again, locking the modes $q_0 \pm 2$ to the central mode q_0 in phase with a phase difference 2ϕ. This scheme will go on until the shifted modes are outside of the amplification spectrum, as shown in Fig. 4.17. Therefore, an inhomogeneously broadened laser medium has to be used that provides gain for all the different longitudinal modes within its amplification spectrum.

To see that these locked modes correspond to a train of short pulses, we investigate the electric field of the laser emission [4]. For simplicity we assume that the locked modes are symmetrically distributed around the central mode q_0 and that they all have the same amplitude E_0. The electric field is then directly given by

$$E(t) = E_0 \sum_{k=-m}^{m} e^{2\pi i [(\nu_0 + k \Delta \nu_{FSR}) t + k\phi]}. \tag{4.79}$$

As the cavity of length L is usually long compared with the length of the laser medium, the free spectral range can be approximated by

$$\nu_{FSR} = \frac{c}{2L}. \tag{4.80}$$

The summation in Eq. (4.79) can be performed analytically, resulting in an electric field

$$E(t) = A(t) e^{2\pi i \nu_0 t}, \tag{4.81}$$

with a time dependent amplitude

$$A(t) = E_0 \frac{\sin\left[(2m+1)\frac{2\pi \Delta \nu_{FSR} t + \phi}{2}\right]}{\sin\left[\frac{2\pi \Delta \nu_{FSR} t + \phi}{2}\right]}. \tag{4.82}$$

Fig. 4.17 Build-up of the longitudinal mode spectrum in a mode-locked laser after the laser emission started on the maximum gain line

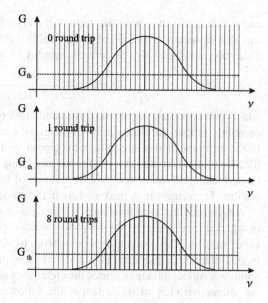

Fig. 4.18 Temporal pulse shape of phase-locked modes for two different values of m

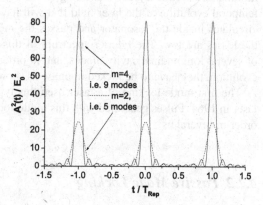

The laser output intensity $I(t) \propto A^2(t)$ will therefore show an amplitude envelope on the high-frequency carrier oscillation ν_0 that corresponds to a train of pulses of width τ_p and a repetition period T_{Rep}, that can be seen in Fig. 4.18. The form of Eq. (4.82) is well known from a multi-slit interference experiment, in which the waves of the evenly spaced slits interfere after a certain distance on a screen. Here, this interference is not an interference in space, but in time, and the different slits correspond to the longitudinal modes that have a evenly distributed phase. The pulse maxima then occur at the times when the denominator in Eq. (4.82) is zero, which corresponds to a repetition rate

$$T_{Rep} = \frac{1}{\Delta \nu_{FSR}} = \frac{2L}{c}, \tag{4.83}$$

and which is just the round-trip time of the laser cavity. Therefore, this pulse train can also be seen as a single pulse with pulse width τ_p that circulates in the cavity. This pulse width can also be derived from Eq. (4.82), resulting in [4]

$$\tau_p \simeq \frac{1}{(2m+1)\Delta\nu_{FSR}}, \tag{4.84}$$

which will approach the inverse gain bandwidth of the laser medium under strong pumping, as then all modes may oscillate. As a result of the temporal interference of the different modes, the pulse peak power will be $(2m+1)^2$ times higher than for a laser in which the same modes are oscillating in an uncorrelated manner. Thus, mode-locking allows not only the generation of very short pulses, but also the generation of extremely high peak powers in the output beam.

An equivalent way of looking at the generation of mode-locked pulses is the case in which a loss is used as the frequency modulator in the cavity near to one of the cavity mirrors. This modulator is then driven by an external signal that causes a loss modulation with a frequency identical to the longitudinal mode spacing $\Delta\nu_{FSR}$. This amplitude modulation now causes the creation of sidebands, as discussed previously. An alternative view of this effect is the following: as the modulator causes loss minima at a frequency corresponding to the round-trip time of the resonator, the temporal evolution of the laser field that will have lowest loss is a short pulse that circulates inside the resonator and passes the modulator just at those times when the losses are low. The Fourier spectrum of this pulse can of-course only consist of several longitudinal cavity modes, and in order to create the pulse-like temporal evolution, they have to be locked in phase as shown in Eq. (4.79).

The first mode-locking of a laser used this type of loss modulation in a He-Ne laser in 1964. Pulses generated with this mode-locking technique usually are on the order of several ps.

4.2.2 Passive Mode Locking

As in passive Q-switching, the use of a saturable absorber in a laser cavity can also cause mode-locking. Therefore, the saturable absorber is placed just in front of the cavity end mirror. When the laser medium now is pumped the laser flux will start spiking as soon as the threshold of the cavity, including the absorber, is reached. This first intensity spike, which will circulate in the cavity with the round-trip time, saturates the absorber more than all other fluctuations of the growing laser field. It will, therefore, see the lowest round-trip loss and has thus maximum amplification and growth rate. As soon as this growing pulse dominates the inversion reduction, the laser will oscillate on a pulse train, which can again be described by Eq. (4.79). However, this point of operation, at which the saturable absorber has enough absorption to favor only one strong noise spike, can be difficult to achieve.

In a temporal scheme the saturable absorber will shorten the rise time of an incident pulse owing to the increasing transmission with increasing pulse intensity. The amplifying laser medium itself creates the opposite process: as a result of the extraction of energy, and thus the reduction in gain, it will shorten the pulse by shortening the fall time of the pulse. However, in most solid-state lasers this effect is low compared with the shortening on the leading edge caused by the saturable absorber, as the upper-state lifetime is usually several orders of magnitude longer than the cavity round-trip time. The only case where both effects are dominant, is in the case of dye lasers, which show excitation lifetimes of the order of the cavity round-trip time. Therefore, ultra-short pulses on the fs-scale were mostly generated with dye lasers in the past.

Using special semiconductor quantum-well structures as saturable absorber and cavity mirror in one (SESAM), which exhibit a strong non-linear response, ultra-short pulses can also created with solid-state lasers. In this case, often a second mode-locking technique is used at the same time, to shorten the pulses further: this is Kerr-lens mode-locking.

Kerr-Lens Mode-Locking

A special way of passive mode-locking is Kerr-lens mode-locking (KLM), in which the self-focusing of an intense laser beam inside an optical medium is used. This effect is based on the Kerr-effect, the increase of the refractive index with increasing intensity $n(I) = n_0 + n_2 I$, and has a response time on the order of fs. In Kerr-lens mode-locking the laser medium is often used as the Kerr medium and an aperture is introduced into the cavity, either by insertion of a solid aperture or by a soft aperture, i.e. by the pumped volume. Assuming a parabolic intensity distribution inside the Kerr medium and a focal length much longer than the Kerr medium itself, it can be shown that the focal length of the Kerr lens is approximatively given by [3]

$$f_{Kerr} \approx \frac{w^2}{4n_2 I_0 L},$$ (4.85)

where, I_0 is the laser peak intensity, L the length of the Kerr medium and w the beam radius inside the Kerr medium. For a Ti:sapphire laser rod of $L = 4$ mm ($n_2 = 3.45 \times 10^{-16} \frac{cm^2}{W}$) and a 200 kW peak power beam focused to $w = 50$ μm, i.e. a peak intensity of

$$I_0 = \frac{P}{\pi w^2} = 2.5 \frac{GW}{cm^2},$$ (4.86)

we obtain a focal length of $f_{Kerr} \approx 18$ cm.

A Gaussian intensity distribution, e.g., thus exhibits a higher refractive index in its center compared with the wings of the radial intensity distribution. Hence, the Kerr medium acts as a positive lens and will focus the beam. Owing to the short response time, the strength of this focusing will be time dependent and only the temporally inner part of a laser pulse will see low losses at the aperture, as shown in Fig. 4.19. The leading and falling edge will be cut off, as their intensity is not sufficient to focus the beam through the aperture with low losses. In the case of a

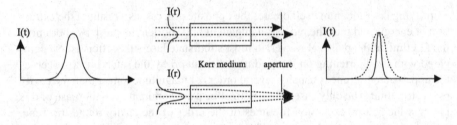

Fig. 4.19 Kerr-lens mode-locking

soft aperture, the focusing will increase the overlap between the beam and the pump volume, thus creating a higher gain for the high-intensity parts of the pulse, which also shortens the pulse. Using KLM in combination with a SESAM, it was possible to generate pulses of ~ 6.5 fs from a Ti:sapphire laser, a solid-state laser with a large gain bandwidth.

4.2.3 Pulse Compression of Ultra-Short Pulses

As already mentioned in Sect. 4.2, short pulses can exhibit a chirp. To understand how this chirp can build-up and how pulses can be compressed by reducing this chirp will be discussed in the following section. Therefore, we investigate the evolution of an incident laser pulse with an electric field amplitude of

$$E_i(t) = E_0 e^{-\xi_0 t^2} e^{i\omega_0 t} \tag{4.87}$$

propagating in a dispersive medium, where, the Gaussian parameter of the incident pulse is given by

$$\xi_0 = a_0 - i b_0. \tag{4.88}$$

The spectrum of this pulse is then expressed by

$$\tilde{E}_i(\omega) = \tilde{E}_0 e^{-\frac{(\omega-\omega_0)^2}{4\xi_0}}. \tag{4.89}$$

In a dispersive medium, the propagation constant $\beta(\omega)$ will show a non-linear dependence on ω, and can thus be approximated around the center frequency ω_0 by

$$\beta(\omega) \approx \beta_0 + \beta_1(\omega - \omega_0) + \beta_2(\omega - \omega_0)^2. \tag{4.90}$$

Thus, the spectrum of the pulse will change during the propagation according to

$$\tilde{E}(\omega, z) = \tilde{E}_i(\omega) e^{-i\beta(\omega)z}. \tag{4.91}$$

Using the Fourier transformation, this corresponds to a time dependence of the electric field of

$$E(t, z) = E_0 e^{i(\omega_0 t - \beta_0 z)} e^{-\xi(z)(t - \beta_1 z)^2}, \tag{4.92}$$

where, $\xi(z)$ is given by

$$\frac{1}{\xi(z)} = \frac{1}{\xi_0} + 2i\beta_2 z. \tag{4.93}$$

From Eq. (4.92) it can be seen that β_0 causes a propagation-distance-dependent phase delay as for any plane wave in a medium with an effective refractive index $n_{eff} > 1$, which can be expressed in terms of the **phase velocity**

$$v_{ph} = \frac{\omega_0}{\beta_0}. \tag{4.94}$$

In an optical medium with refractive index n, the phase velocity is given by $v_{ph} = \frac{c}{n}$ and $\beta_0 = k_z$ corresponds to the wave vector component in the propagation direction. However, in optical waveguides such as optical fibers the dispersion and index properties of the medium are changed resulting from the wave-guiding effect.

The influence of β_1 affects the Gaussian envelope of the electric field by introducing a delay on the envelope, which now propagates with the so-called **group velocity**

$$v_g = \left(\frac{\partial\beta}{\partial\omega}\right)^{-1}\Bigg|_{\omega=\omega_0} = \frac{1}{\beta_1}, \tag{4.95}$$

and the effect of β_2 is a change in the Gaussian parameter $\xi(z)$ with propagation distance, thus changing the shape of the pulse envelope, i.e. its pulse width and the chirp. As β_2 can be expressed as

$$\beta_2 = \left[\frac{\partial}{\partial\omega}\left(\frac{1}{v_g(\omega)}\right)\right]_{\omega=\omega_0}, \tag{4.96}$$

it is also called **group-velocity dispersion**. This influence on the pulse can be derived from Eq. (4.93), from which the real and imaginary part of the Gaussian parameter $\xi(z) = a(z) - ib(z)$ can be deduced as

$$a(z) = \frac{a_0}{(1 + 2\beta_2 b_0 z)^2 + (2\beta_2 a_0 z)^2}, \tag{4.97}$$

$$b(z) = \frac{b_0 + 2\beta_2 z(a_0^2 + b_0^2)}{(1 + 2\beta_2 b_0 z)^2 + (2\beta_2 a_0 z)^2}. \tag{4.98}$$

From Eqs. (4.97), (4.97), we can deduce why ultra-short pulses usually exhibit a chirp. Assuming a chirp-free Gaussian pulse, i.e. $b_0 = 0$, we find that by propagating this pulse in a dispersive medium, e.g. an output coupler mirror substrate, vacuum windows or an optical fiber with non-zero group-velocity dispersion, it will exhibit an increasing chirp, which after a propagation length z in this medium is given by

$$b(z) = \frac{2\beta_2 z a_0^2}{1 + (2\beta_2 a_0 z)^2} = \frac{1}{2\beta_2}\frac{z}{z^2 + (\frac{\tau_p^2}{4\beta_2 \ln 2})^2}. \tag{4.99}$$

This chirp build-up is shown in Fig. 4.20, where, the reference length z_0 is given by

$$z_0 = \frac{\tau_p^2}{4\beta_2 \ln 2}. \tag{4.100}$$

Fig. 4.20 Increasing chirp of an unchirped pulse propagating in a medium with group-velocity dispersion

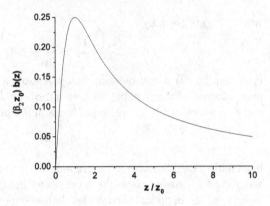

Fig. 4.21 Increasing pulse width of an unchirped pulse propagating in a medium with group-velocity dispersion

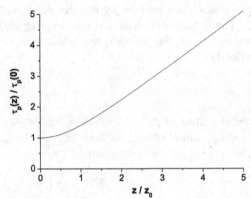

The pulse width then results in

$$\tau_p(z) = \tau_p(0)\sqrt{1 + \left(\frac{z}{z_0}\right)^2}, \tag{4.101}$$

a relation equivalent to the evolution of a the radial width of a Gaussian beam, see Eq. (3.44). Thus the incident pulse width will increase with propagation distance, as shown in Fig. 4.21.

However, as can also be seen from Eqs. (4.97), (4.98), a chirped pulse with incident chirp $b_0 \neq 0$ can be compressed in pulse width if a medium with proper group-velocity dispersion is used. The optimum group-velocity dispersion interaction length is given by

$$2\beta_2 L_{opt} = -\frac{b_0}{a_0^2 + b_0^2}, \tag{4.102}$$

which results in a maximum of $a(z)$ at $b(z) = 0$, and therefore, in a minumum pulse width

$$\tau_{p,min} = \frac{\tau_p(0)}{\sqrt{1 + (\frac{b_0}{a_0})^2}}. \tag{4.103}$$

Fig. 4.22 Pulse compression
using a pair of diffraction
gratings

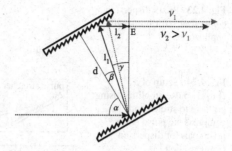

This corresponds to a pulse from which all chirp has been removed and transformed
into its short pulse width. A large chirp will thus yield a high pulse width compression ratio with a final pulse that is Fourier transform limited when all chirp has been
removed, as can be seen by inserting this result into Eq. (4.78).

Pulse Compression Methods

Depending on the actual chirp of the pulse, a medium or optical system with the
proper group-velocity dispersion, is necessary in order to compress the pulse. Instead of using a massive medium with its natural dispersion, optical systems consisting of gratings or prisms are mostly used in these laser designs.

The grating design uses two diffraction gratings as shown in Fig. 4.22. Owing
to the wavelength dependent diffraction angle, the internal optical path length of
this grating system differs for different wavelengths, and will consequently create
the necessary wavelength-dependent time delay to compress the pulse. Use of this
technique enables large dispersion effects can be generated to compress pulses with
strong chirps. The path length ΔL between the common incident point on the first
grating and the point on the common exit plane of the grating compressor E, is
given by

$$\Delta L = l_1 + l_2 = \frac{d}{\cos \beta} + \frac{d}{\cos \beta} \sin \gamma = \frac{d}{\cos \beta}(1 + \sin \gamma). \qquad (4.104)$$

Using the grating equation

$$\frac{\lambda}{g} = \sin \alpha - \sin \beta, \qquad (4.105)$$

in which g is the grating period, the spatial dispersion is given by

$$\frac{\partial \Delta L}{\partial \lambda} = \frac{\partial \Delta L}{\partial \beta} \frac{\partial \beta}{\partial \lambda} = \frac{\lambda d}{g^2 \cos^3 \beta} = \frac{\lambda d}{g^2 (1 - (\sin \alpha - \frac{\lambda}{g})^2)^{\frac{3}{2}}}. \qquad (4.106)$$

Therefore, the internal path length of the grating compressor will increase with
wavelength, and will thus create a larger time delay between the entry and exit
planes for larger wavelengths. Hence, it will compress a pulse with a positive chirp.

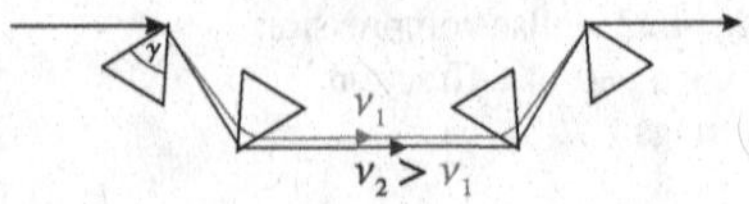

Fig. 4.23 Prism dispersion compensator for intracavity applications

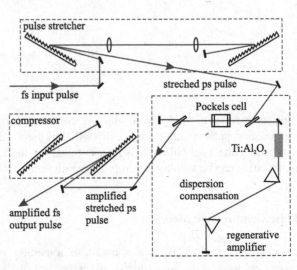

Fig. 4.24 Setup of a chirped-pulse amplifier, using gratings to stretch and compress the pulse and a prism pair for dispersion compensation [3]

A second alternative, which is mostly used to introduce a small correction to the cavity dispersion in a laser resonator for short-pulse mode-locking, is based on a system of prisms, which again shows a wavelength-dependent optical path length. In this case the prism material, as well as the geometry (prism angle γ), can be chosen so that the laser beam is incident on to the prism surfaces at Brewster's angle, for which the reflection losses are greatly reduced. The design in Fig. 4.23 also allows an easy insertion into an existing resonator, as the input and output beams are colinear. The strength of the dispersion introduced by this system, is lower than for a grating compressor; however, this prism compressor can be used to generate both signs of dispersion, i.e. either $\beta_2 < 0$ or $\beta_2 > 0$.

Chirped-Pulse Amplification

A main application of pulse compressors and its counterparts, pulse stretchers, is found in chirped-pulse amplification. This technique, depicted in Fig. 4.24, allows the generation of high-pulse-energy fs pulses. First, a standard fs mode-locked laser oscillator is used to generate fs pulses at a repetition rate of around 80 MHz with pulse energies on the order of some nJ. The oscillator pulses are then stretched in pulse width by passing an anti-parallel grating pair including a 1 : 1 telescope, creating a strong chirp on the pulse. Consequently, the pulse width is increased, e.g. by a factor of 3000 from 200 fs to 600 ps, decreasing enormously the pulse peak power. This now allows a high amplification of the pulses to pulse energies of several mJ without reaching the optical damage thresholds of the components in the amplifier.

The amplifier setup shown in Fig. 4.24 is a **regenerative amplifier**. The input pulses enter the amplifier by an intracavity polarizer. Then one pulse is selected by switching the Pockels cell to its half-wave voltage to rotate the polarisation of this pulse by 90°. It now passes the second polarizer and gets reflected by the cavity end mirror. At that time the Pockels cell is switched off. Thus the pulse keeps its polarisation and resonates back and forth in the amplifier cavity, where it passes a Ti:sapphire laser gain element twice for each round-trip.

In order to compensate for the internal cavity dispersion, a prism pair is also inserted into the amplifier cavity. When the pulse has made sufficient round-trips to reach its maximum pulse energy, the Pockels cell is switched to its half-wave voltage again while the pulse travels on the prism side of the cavity. Thus, when it comes back to the Pockels cell, it will be rotated by 90° and leaves the cavity by the second polarizer. Finally, the pulse chirp is removed in a grating compressor, reducing the pulse width back to the fs scale of the input pulse. Usually, the final pulse width is a bit longer than the original pulse width of the input pulse as a result of some additional higher-order chirp accumulated during the amplification steps. Whilst the damage threshold of the optical components in the stretched part of the setup are usually high enough, a critical point is the final grating of the compressor, at which the high-energy pulse has been compressed to its short pulse width, resulting in extreme peak powers. In order to prevent damage on that gratings, the beam diameter has to be strongly increased, making large-aperture gratings necessary.

References

1. A.E. Siegman, *Lasers* (University Science Books, Sausalito, 1986)
2. L.M. Frantz, J.S. Nodvik, J. Appl. Phys. **34**, 2346 (1963)
3. W. Koechner, *Solid-State Laser Engineering* (Springer, Berlin, 1999)
4. F.K. Kneubühl, M.W. Sigrist, *Laser* (Teubner, Stuttgart, 1999)

Chapter 5
Laser Examples and Their Applications

In this chapter we will investigate different types of practical lasers that are often used in the laboratory. Owing to the recent advances in high-power and high-brightness laser diodes, diode-pumped solid-state lasers are the most important lasers today and into the future. Therefore, the lasers described below all belong to the solid-state laser category, with the exception of the HeNe laser, a gas laser, which is a well known device still widely used as precision alignment laser in the laboratory. Another important gas laser is the CO_2 laser, which uses a molecular transition excited via electronically excited N_2 molecules, using a He buffer gas for cooling and reduction of the lifetime of the lower laser level. This laser is the most efficient gas laser with electrical-to-optical efficiencies of up to 30 %.

Other laser types are, e.g. dye lasers, in which a dye solution is used as the active medium. These lasers were often used to provide wide-band tunable sources from the ultra-violet to the infrared spectral region or to generate ultra-short pulses, as the missing long-range order in the liquid laser medium results in broad transitions.

Also lasers without an active medium exist, the free-electron lasers. In these devices, a relativistic electron beam is sent into an alternating-pole spatially-periodic magnetic field, which forces the electrons on to an undulating path. As these electrons are accelerated charges, they emit a synchrotron radiation with a wavelength that depends on the period of the magnets, the relativistic contraction of this period in the frame of the electron beam and the relativistic Doppler shift back to the laboratory frame.

5.1 Gas Lasers: The Helium-Neon-Laser

The HeNe laser was the first cw laser and also the first gas laser realized. While at that time the strongest transition at 1.15 µm was used, the HeNe lasers today are mainly operated on the visible lines, with the mostly used line at 632.8 nm in the red spectrum. Other important lines are the green 543.3 nm, the yellow 594.1 nm and the orange 611.8 nm lines. Other infrared lines are the near-infrared lines at 1152.3 nm and 1523.1 nm as well as the 3391.3 nm mid-infrared line.

M. Eichhorn, *Laser Physics*, Graduate Texts in Physics,
DOI 10.1007/978-3-319-05128-4_5,
© Springer International Publishing Switzerland 2014

Fig. 5.1 Energy scheme of
the HeNe laser including the
main transitions

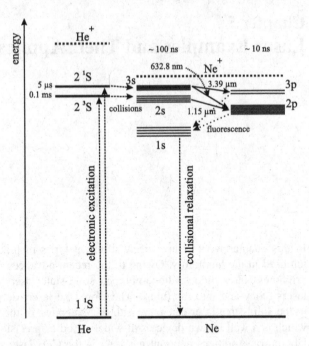

In the HeNe laser [1] the laser active medium is the neon gas. The addition of helium is only used for the pumping process. It also contributes to the cooling of the gas mixture due to the high thermal conductivity of He. Usually, a mixture of about 0.1 mbar Ne in 1 mbar He is used. As can be seen in Fig. 5.1, the pumping of the Ne atoms occurs in an electric discharge in a two-stage process: excitation of He and energy transfer to Ne. First the He atoms are excited by collisions with the electrons in the discharge, bringing them into the metastable levels $2\,^3S$ and $2\,^1S$, with lifetimes of 0.1 ms and 5 μs, respectively. Then, owing to the nearly coincidence between these levels and the $2s$ and $3s$ levels of Ne, the stored energy will be transferred to the Ne atoms in atomic collisions between He and Ne. As the lifetime of the $2s$ and $3s$ levels of Ne is on the order of 100 ns, a population inversion results with respect to the $2p$ and $3p$ levels, which exhibit a lifetime of only around 10 ns. As a result of the selection rules of electric dipole transitions the Ne atoms can only emit on lines connecting a s and a p state, resulting in the above-mentioned laser transitions. From the p states the Ne ions quickly relax to the $1s$ state by fluorescence emission. As this state is also metastable, i.e. long lived, the Ne ions would be re-excited into the $2p$ state by electron collisions, where they would cause a re-absorption on the laser lines terminating in this state. To avoid a strong population of this $1s$ state, small bore discharge tubes are used to cause a decay of this state back to the ground state by collisions with the wall of the laser tube.

Resulting from the dependence of the emission cross-section

$$\sigma_e(\nu_s) \propto \frac{g(\nu_s)}{\nu_s^2} \qquad (5.1)$$

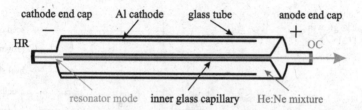

Fig. 5.2 Cut through a typical HeNe laser tube. Here, the resonator mirrors are bonded to the glass tube directly in "hardseal" technology

on the laser frequency ν_s, as shown in Eq. (1.74), and the dependence of the dominating Doppler broadening line form factor from Eq. (1.84), given by

$$g(\nu_s) \propto \frac{1}{\nu_s},\qquad\qquad(5.2)$$

the maximum gain of the HeNe laser will be proportional to λ_s^3 and will thus occur on the mid-infrared transition at 3391.3 nm. Therefore, this line will usually show the lowest pump threshold and the HeNe laser would only emit at this line. In order to avoid this effect and to operate the laser on the other lines, special cavity mirrors are used that do not reflect the mid-infrared radiation, thus greatly increasing the threshold power at this line well above the thresholds of the visible lines. A second possibility is to insert a quartz-glass plate, preferentially at Brewster's angle, into the cavity. This strongly increases the intra-cavity losses on the mid-infrared line owing to the internal infrared absorption of the glass whilst nearly no losses are added for the visible-to-near-infrared transitions. Additionally, the laser output will be linearly polarized, as only the s-polarization of the cavity mode will be transmitted through the Brewster plate without Fresnel reflection loss.

The experimental construction scheme of a HeNe laser is shown in Fig. 5.2. The geometry of the capillary has to be chosen in a way so that the product of total gas pressure p and capillary bore diameter d is about $pd \simeq 4.8$–5.3 mm mbar, whilst the optimum mixture between He and Ne depends on the emission line. For the 632.8 nm line, a partial pressure ratio of He : Ne = 5 : 1 is used, whereas an optimum ratio of He : Ne = 9 : 1 was found for the 1152.3 nm line. A third parameter is the discharge current density, which is especially important for the 632.8 nm and 3391.3 nm lines.

The main applications of HeNe lasers today are as alignment sources with high beam quality, caused by the low beam distortions generated in a gas laser, and as highly coherent laser sources in holography, interferometry and ring-laser gyroscopes. As a result of the fact that the laser transitions occur between very high-energetic levels, as can be seen in Fig. 5.1, the quantum efficiency of the HeNe laser is around 10 %. However, the total electrical-to-optical efficiency of the HeNe laser is very low, usually around 0.1 %, which is caused by the low efficiency excitation mechanism within the plasma discharge [1].

Fig. 5.3 Schematic setup of
a Czochralski growth
apparatus [12]

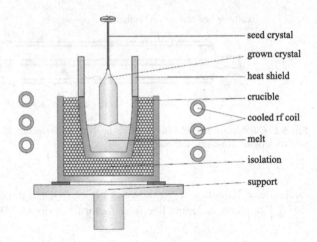

seed crystal

grown crystal

heat shield

crucible

cooled rf coil

melt

isolation

support

5.2 Solid-State Lasers

The most important lasers today are solid-state lasers, in which ionic impurities
doped into transparent crystals or glasses act as the laser active species. In general,
two main types of hosts and two main types of active ions can be distinguished:
crystals and glasses on the one side, and rare-earth ions and transition-metal ions on
the other side.

In rare-earth ions, the laser active electronic states are located in the inner $4f$
shell of the ion. Therefore, they are shielded to a great extent from the crystal field
by the outer shell electrons ($5s^2$ and $5p^6$ electrons). Thus, the crystal field and the
coupling of the ionic states to the phonons of the host lattice is low, resulting in
usually narrow line widths for the optical transitions. In the spectra of the rare-earth
ions in crystalline media, the different lines of the transitions between the various
Stark levels are clearly visible as already shown in Fig. 2.6.

If these rare-earth ions are doped into a glass matrix, the arguments concerning
the influence of the crystal field on to the transitions is still valid. However, the glass
is an amorphous solid and the geometric structure of the glass matrix, and thus the
crystal field varies locally, causing a spatially dependent line shift. This causes an
inhomogeneous broadening of the emission spectrum and results in a very broad
gain of these laser media. This is important to make broadly tuneable lasers, as well
as for the generation of ultra-short pulses.

In contrast to the rare-earth ions, the optically active electrons in transition-metal
ions are located in the outer shells of the ion and are thus fully affected by the crystal
field. They, therefore, show a very strong coupling to the lattice phonons in crystals,
which result in mixed electronic-vibronic states. Owing to this effect the widths of
the transition lines are also extremely broad, making e.g. the Ti:sapphire laser so
important for the generation of ultra-short pulses.

Crystal Growth   Most of the laser crystals used today are grown by the Czochralski method shown in Fig. 5.3. This technique uses a single-crystalline seed, which

Fig. 5.4 Atomic structure of the YAG crystal and the sites of the Nd^{3+} ion in YAG [13]

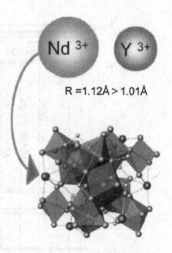

is used to pull a large single crystal from a melt, consisting of a stoichiometric composition of primary chemicals. In the case of the well-known laser host yttrium-aluminum-garnet (YAG, $Y_3Al_5O_{12}$), this is a mixture of yttrium oxide and aluminum oxide, to which a small amount of rare-earth oxide, e.g. neodymium oxide, is added, determining the rare-earth dopant concentration of the final laser crystal. The composition is melted in a crucible consisting of a metal of high melting point, such as iridium, which is heated by induction from an external, water-cooled rf coil. The crucible itself is embedded with thermally insulating pellets, such as zirconia. A single crystalline seed is mounted on a rotating rod and brought into contact with the melt. Then, this rotating rod is slowly pulled with a speed of some mm per hour, causing the single crystal to grow. When the growth process is completed, the finished crystal is slowly cooled down to room temperature to anneal internal stress that may have built up during growth.

5.2.1 The Nd^{3+}-Laser

The neodymium laser is one of the mostly used lasers today, operating often on a true four-level transition with an emission wavelength of 1064 nm for Nd^{3+} in YAG as the host crystal. This special host crystal is shown in Fig. 5.4. Where, the Nd^{3+} ion replaces the Y^{3+} ion. However, owing to the larger ionic radius of Nd^{3+} compared with Y^{3+}, only a small fraction of the Nd^{3+} ions are incorporated into the crystal structure during growth, resulting in Nd^{3+} dopant concentrations of usually 0.1–1.2 %, as well as a doping gradient along the growth direction. This gradient is caused by the increased concentration of the Nd^{3+} ions in the melt during growth, resulting in an increase in Nd^{3+} dopant concentration in the crystal towards the end of the growth process.

The energy level scheme of Nd^{3+} ions in different hosts is shown in Fig. 5.5. The upper laser level lifetime in YAG is 250 µs. It is the 1064 nm transition, which

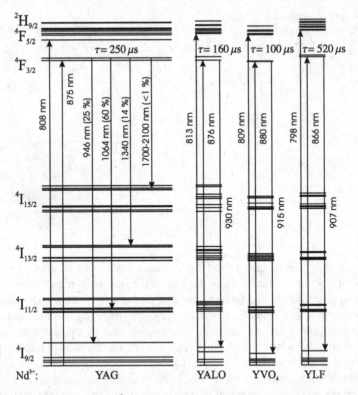

Fig. 5.5 Energy scheme of the Nd^{3+} ion in various hosts and the main pump and laser transitions

shows the largest fluorescence, to which 60 % of all radiatively decaying ions in the $^4F_{3/2}$ manifold contribute to. The fluorescence corresponding to the quasi-three-level transition around 900–950 nm is caused by 25 % of the decaying ions and the second four-level transition at around 1.34 μm arises from a contribution of 14 % of the total decay. The mid-infrared transitions around 2 μm are so weak that they are not used for any practical laser. Owing to the relatively large emission cross section of the 1064 nm transition compared with the 946 nm and the 1342 nm transition, special care has to be taken whenever these two weaker transitions are to be used. In this case, special cavity mirrors and laser media, which are anti-reflection coated at 1064 nm have to be used in order to suppress any feedback on the 1064 nm line. Otherwise, the laser would oscillate on this 1064 nm line, making emission on the other lines impossible by fixing the upper level population to a value that does not allow the threshold gain for the other transitions to be achieved.

The Nd^{3+} ion is either pumped by flashlamps or more recently by high-power laser diodes, which can be directly designed to match the absorption spectrum of the two most important pump transitions at around 808 nm and 875 nm. As a result of this absorption-matched pumping laser-diode-pumped Nd^{3+} lasers show high efficiencies $\gg$40 %, whilst only a small part of the full emission spectrum of a flashlamp will be absorbed by the Nd^{3+} ion, resulting in a total efficiency of usu-

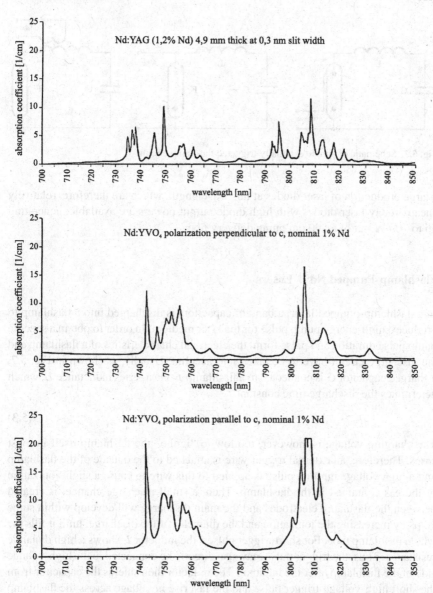

Fig. 5.6 Pump absorption coefficients for Nd^{3+} in different hosts. (Source: Northrop Grumman Corporation, USA)

ally <1 % in flashlamp-pumped lasers. Therefore, flashlamp-pumped Nd^{3+} laser crystals are often co-doped with Cr^{3+} ions, which show a large absorption for the flashlamp emission spectrum. The energy absorbed by the Cr^{3+} ions is then transferred to the Nd^{3+} ions in a direct ion-ion energy transfer process. Diode pumping at around 808 nm on the absorption lines shown in Fig. 5.6 is very popular and led to

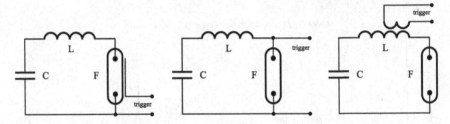

Fig. 5.7 Schematic of simple externally-triggered flashlamp circuits

a large production of laser diodes at this wavelength, which are therefore, relatively cheap. Today laser diodes with high diode output powers are available, generating up to 100 W out of a 200 μm pump delivery fiber.

Flashlamp-Pumped Nd^{3+} Lasers

In a flashlamp-pumped laser, a bank of capacitors is discharged into a flashlamp to produce a high-energy pump pulse for the laser medium. In order to obtain a specific pump pulse duration and pulse form, the electrical characteristics of a flashlamp and the discharge circuit have to be taken into account. Usually, as shown in Fig. 5.7, a charged capacitor C is connected to a flashlamp F via a series inductance L, which determines the discharge time constant

$$T = \sqrt{LC}. \tag{5.3}$$

The charging voltage is, however, too low to "ignite" the flashlamp itself in most cases. Therefore, an external trigger wire is attached to the outside of the flashlamp and a high voltage ignition pulse is applied to this wire to cause a slight ionization of the gas contained in the flashlamp. Then, a small discharge channel is created between the flashlamp electrodes and the main discharge will develop within some 10 μs by increasing the ionization and the diameter of the discharge, until it fills the whole flashlamp tube. For short trigger pulses, the inductor L shows a high dynamic resistance. Thus, the trigger pulse may also directly be connected to the high-voltage side of the flashlamp after the inductor. The inductor then shields the capacitor from the short high-voltage-trigger pulse and the fast rise in voltage across the flashlamp causes breakdown of the gas in the flashlamp. A third possible trigger circuit uses the inductor itself as a secondary of a transformer, and the trigger pulse is applied to the primary. Thus, the trigger pulse will induce a high voltage in the secondary, which adds to the voltage of the capacitor and causes breakdown of the flashlamp.

After the discharge has fully developed and fills the whole flashlamp tube, the flashlamp shows a non-linear resistance and the voltage across the flashlamp U is connected to its current I by [3]

$$U = K_0\sqrt{I}, \tag{5.4}$$

Fig. 5.8 Schematic of a
standard RLC circuit

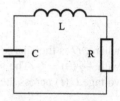

where, K_0 is a parameter of the flashlamp, which is given by the manufacturer or can be measured experimentally. It depends on the geometry of the flashlamp, i.e. its arc length l and the inner tube diameter d, and the gas parameters as

$$K_0 = k\frac{l}{d}. \tag{5.5}$$

For 450 torr xenon-filled flashlamps, a value of $k = 1.27\ \Omega\sqrt{A}$ is found. Including the dependence on the gas pressure, the flashlamp constant can be described by

$$K_0(P) = K_0(P_0)\left(\frac{P}{P_0}\right)^{\frac{1}{5}} \tag{5.6}$$

for xenon-filled flashlamps. The reference pressure in this case is $P_0 = 450$ torr. For krypton, approximately the same flashlamp constant is found [3]. Thus, the electric resistance of the flashlamp can be described by

$$R_F = \frac{K_0}{\sqrt{I}} \tag{5.7}$$

when the discharge has fully developed.

Another important parameter of a flashlamp is its explosion energy, which gives the amount of electrical input energy that will cause catastrophic damage to the tube wall. This damage is caused by the high temperature of the plasma and especially the acoustic shock wave generated by the plasma, which itself heats up during the pulse from 300 K to about 12000 K. The explosion energy is related to the surface of the inner tube wall ld and the duration of the pulse t_p by

$$E_X = k_X ld\sqrt{t_p}, \tag{5.8}$$

where, k_X is a parameter that depends on the gas filling and gas pressure. Using the definition of the explosion energy the lamp life, i.e. the number of shots N a flashlamp can usually be used at an electrical pulse energy per shot of E_0, can be derived empirically and is related to its single-shot explosion energy by

$$N \approx \left(\frac{E_X}{E_0}\right)^{8.5}. \tag{5.9}$$

Therefore, the lamps usually are operated well below their explosion energy, resulting in a nominal lifetime of 10^6–10^8 shots.

In the case of a constant load resistance R instead of a flashlamp, a standard RLC circuit results as shown in Fig. 5.8. The differential equation governing the discharge evolution of the capacitor in this case is given by

$$L\frac{\partial^2 Q}{\partial t^2} + R\frac{\partial Q}{\partial t} + \frac{Q}{C} = 0, \tag{5.10}$$

where, Q is the charge of the capacitor C and the initial condition is $Q(0) = CU_0$ with $\frac{\partial Q}{\partial t}(0) = 0$. The solution of this equation can be easily found, resulting in a voltage $U(t)$ across the resistor R of

$$U(t) = U_0 \frac{\gamma}{\omega} e^{-\gamma t} \left(e^{i\omega t} - e^{-i\omega t} \right), \tag{5.11}$$

with

$$\gamma = \frac{R}{2L} \tag{5.12}$$

$$\omega = \sqrt{\frac{1}{LC} - \gamma^2}. \tag{5.13}$$

The circuit thus shows three different cases of discharge behaviour, depending on the actual values of the components:

• **Underdamped discharge.** Here

$$R < 2\sqrt{\frac{L}{C}}, \tag{5.14}$$

resulting in a real value of ω. Thus, the current will show an oscillatory behaviour, which is exponentially damped due to the energy dissipation in the resistance R.

• **Overdamped discharge.** Here

$$R > 2\sqrt{\frac{L}{C}}, \tag{5.15}$$

resulting in an imaginary value of ω. Therefore, no oscillatory current will build up. The high resistance results in a low peak current and it will take a long time until the capacitor is fully discharged. Both of these operation regimes are usually not desired in flashlamp circuits for lasers. The oscillatory discharge causes an erosion of the flashlamp electrodes, which are designed for a specific polarity, and the overdamped discharge results in low pump peak intensities.

• **Critically damped discharge.** In this special case

$$R = 2\sqrt{\frac{L}{C}}, \tag{5.16}$$

i.e. $\omega = 0$, the current will not show an oscillatory behaviour and the stored energy is delivered to the load in the shortest possible time without oscillation. The absolute values of the current and the voltage follow the relation [3]

$$I(t) = I_{peak} \frac{t}{T} e^{-\frac{t}{T}+1}, \tag{5.17}$$

$$U(t) = 2U_0 \frac{t}{T} e^{-\frac{t}{T}}, \tag{5.18}$$

with $I_{peak} = \frac{2U_0}{eR}$ being the peak discharge current and $T = \sqrt{LC}$ being the time constant of the LC circuit. The current pulse form and the corresponding capacitor voltage is shown in Fig. 5.9.

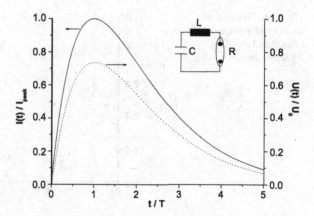

Fig. 5.9 Characteristic current and voltage evolution of a critically-damped discharge circuit

However, in a flashlamp circuit the flashlamp shows a non-linear, current-dependent resistance [3]. Using the definition of the wave resistance

$$Z_0 = \sqrt{\frac{L}{C}}, \qquad (5.19)$$

the damping factor γ for the flashlamp is given by

$$\gamma = \frac{K_0}{T\sqrt{U_0 Z_0}}, \qquad (5.20)$$

and thus, depends on the starting voltage of the capacitor. The critically damped case usually employed in a laser system corresponds to $\gamma = 0.8T^{-1}$, for which a pulse width of $t_p = 3T$ can be deduced, defined as the time between the 10 % points of the current pulse, and corresponding to approximately 97 % of the total discharge energy delivered to the flashlamp. Using the relation of the stored energy E_0 in the capacitor,

$$E_0 = \frac{1}{2}CU_0^2, \qquad (5.21)$$

the necessary capacity is found from (5.20) as

$$C^3 = \frac{2\gamma^4 T^4}{9} \frac{E_0 t_p^2}{K_0^4}, \qquad (5.22)$$

resulting for $\gamma = 0.8T^{-1}$ in

$$C^3 = 0.091 \frac{E_0 t_p^2}{K_0^4}. \qquad (5.23)$$

Hence, the necessary inductance is

$$L = \frac{t_p^2}{9C}. \qquad (5.24)$$

It has to be noted that this calculation gives the design for a critically-damped circuit at a certain lamp energy E_0, and therefore, for a certain capacitor voltage U_0. When

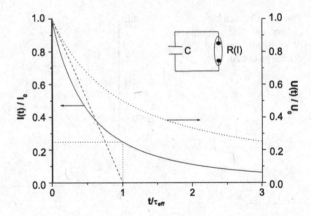

Fig. 5.10 Characteristic current and voltage evolution of an inductance-free discharge circuit

the capacitor is charged to a higher voltage, the system is no longer critically damped and discharge oscillations will occur. For this critically-damped case, it has been found empirically [3] that the explosion energy of a xenon-filled flashlamp can be described by,

$$E_X = 1.2 \times 10^4 \, \mathrm{J\,cm^{-2}\,s^{-1/2}} ld\sqrt{t_p}, \tag{5.25}$$

i.e. a parameter $k_X = 1.2 \times 10^4 \, \mathrm{J\,cm^{-2}\,s^{-1/2}}$.

In some cases, especially for extremely compact lasers and circuits where inductors are not used, then this circuit corresponds to the completely overdamped case. Then, by taking the non-linear resistance of the flashlamp into account, Eq. (5.10) becomes

$$K_0\sqrt{-\frac{\partial Q}{\partial t}} + \frac{Q}{C} = 0, \tag{5.26}$$

where, the current was taken as $I = -\frac{\partial Q}{\partial t}$. We define the effective time constant of the discharge by

$$\tau_{eff} = \frac{K_0^2 C}{U_0}, \tag{5.27}$$

which depends on the charging voltage U_0 of the capacitor. Then, we obtain for the capacitor voltage $U(t)$ and the circuit current $I(t)$

$$U(t) = \frac{U_0}{\frac{t}{\tau_{eff}} + 1}, \tag{5.28}$$

$$I(t) = \frac{I_0}{(\frac{t}{\tau_{eff}} + 1)^2}, \tag{5.29}$$

with a peak current of

$$I_0 = \frac{U_0^2}{K_0^2}. \tag{5.30}$$

The characteristic pulse form is shown in Fig. 5.10. The energy delivered to the flashlamp during the time τ_{eff} is given by

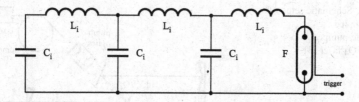

Fig. 5.11 Schematic of a transmission-line flashlamp circuit for rectangular pulses

$$E(\tau_{eff}) = \int_0^{\tau_{eff}} U(t)I(t)dt = \frac{3}{4}E_0. \tag{5.31}$$

After a time of $2\tau_{eff}$, ~89 % of the total energy has been delivered to the flashlamp.

Often, a more rectangular pulse shape is required, especially in free-running lasers, which should emit a rectangular laser pulse. In this case, the total capacity C_{tot} and the total inductance L_{tot} have to be devided as shown in Fig. 5.11, to form a transmission line [3]. Each mesh of this transmission line consists of an LC circuit with $L_i = \frac{L_{tot}}{n}$ and $C_i = \frac{C_{tot}}{n}$, with n being the number of meshes within the transmission line. The characteristic impedance of the transmission line,

$$Z = \sqrt{\frac{L_{tot}}{C_{tot}}}, \tag{5.32}$$

is then chosen to match the load resistance of the flashlamp $R(I) = Z$ at the requisite current. It is convenient to define the pulse width t_p^* in this case, as the time between the 70 % points of the current pulse, resulting in

$$t_p^* = 2\sqrt{L_{tot}C_{tot}} = 2T, \tag{5.33}$$

from which the necessary total capacity

$$C_{tot} = \frac{t_p^*}{2Z} \tag{5.34}$$

and the inductance

$$L_{tot} = \frac{t_p^* Z}{2} \tag{5.35}$$

can directly be calculated. The peak current of the discharge is then given by

$$I_{peak} = \frac{U_0}{2Z}. \tag{5.36}$$

Thus, the charging voltage of the transmission line needs to be set in order to achieve the impedance matching between the transmission line and the current-dependent resistance of the flashlamp. The rise time of the pulse measured between the 10 % and the 80 % point decreases with the number of meshes n as

$$t_r^* = \frac{t_p^*}{2n}. \tag{5.37}$$

Fig. 5.12 Schematic setup of a fiber-coupled laser-diode-pumped Nd^{3+}:YVO$_4$ laser [3]

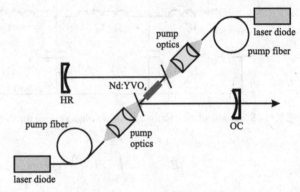

A high number of meshes results in a more and more rectangular pulse. The necessary total capacity can be derived from (5.21), (5.33) and (5.36) and results in

$$C_{tot}^3 = \frac{1}{8} \frac{E_0 t_p^{*2}}{K_0^4}. \tag{5.38}$$

Laser-Diode-Pumped Nd^{3+} Lasers

A Nd^{3+} laser, pumped by fiber-coupled laser diodes, is shown in Fig. 5.12 as an example of a longitudinally pumped Nd^{3+}:YVO$_4$ medium. The cavity is folded by the use of two dichroic mirrors, which are highly reflecting under the chosen angle of incidence for the laser radiation, whilst being highly transmissive for the diode pump beam. The pump output of the delivery fiber is first collimated by a lens and then focused into the crystal with an appropriate focal length in order to get a pump spot that is well matched to the fundamental mode distribution of the cavity inside the laser crystal. The fibers used for high-power pumping are usually multi-mode fibers. Thus, the pump beam propagation can be described by standard geometrical optics. The two main important parameters of the fiber, are its core diameter d and its numerical aperture NA, which is determined by the refractive-index difference in a step-index fiber as

$$NA = \sqrt{n_{core}^2 - n_{cladding}^2}. \tag{5.39}$$

The numerical aperture also describes the half-angle θ_f of the radiation emitted by the fiber, given by

$$NA = \sin\theta_f. \tag{5.40}$$

Usually, the pump optics consist of a two-lens telescope, where, the first lens collimates the beam emitted by the fiber, and the second lens refocuses the beam to form the pump spot in the crystal. When we denote the magnification

$$M = \frac{2r_p}{d} \tag{5.41}$$

of the telescope as the ratio between pump-spot diameter to fiber core diameter, the internal angle of the pump radiation in the crystal can be described by

$$\theta_i = \arcsin\left(\frac{1}{n}\sin\left[\frac{\arcsin NA}{M}\right]\right) \approx \frac{NA}{nM}, \tag{5.42}$$

which can be approximated in most cases for small values of the NA, where, n is the refractive index of the laser crystal. Thus, especially when long laser media are used, the pump beam inside the crystal is no longer a cylindric volume. To describe the laser behaviour in this case, especially the threshold pump power, we have to define an effective pump beam radius $w_{p,eff}$ and thus, an effective pump beam area A_{eff}. To model the laser output power, the already known linear relation with respect to pump power given by

$$P_{out} = \eta_{slope}(P_p - P_{th}) \tag{5.43}$$

is used, with the threshold pump power

$$P_{th} = \frac{I_{sat}^p A_{eff}}{\eta_{abs}}(\ln G + \sigma_a(\lambda_s)\langle N\rangle L) \tag{5.44}$$

and the slope efficiency

$$\eta_{slope} = \eta_{mode}\frac{\lambda_p}{\lambda_s}\frac{-\ln(1-T_{OC})}{2\ln G}\eta_{abs}, \tag{5.45}$$

where, $T_{OC} = 1 - R_{OC}$ is the output coupler transmission,

$$\eta_{abs} = 1 - e^{-\alpha_p L} \tag{5.46}$$

the fraction of absorbed pump power, α_p the pump absorption coefficient, $\langle N\rangle$ the average dopant ion density, I_{sat}^p the pump saturation intensity, G the single-pass gain, and $A = \pi w_{eff}^2$ the effective pump beam area. As pump and laser beam often do not overlap exactly, a mode fill efficiency η_{mode} is introduced into the slope efficiency. To find a description for the effective pump beam radius, we have to take into account the fact that the beam radius will change axially as a result of focusing and that the pump intensity will, in addition, change owing to the absorption along the crystal. The axial evolution of the real pump beam radius can be described by

$$w_p(z) = r_p\sqrt{1 + \left(\frac{z - z_0}{r_p}\tan\theta_i\right)^2}, \tag{5.47}$$

where, r_p denotes the pump beam focal spot radius inside the crystal and z_0 the position of the focus. As the local pump efficiency depends on the local pump intensity, the effective pump beam radius can be described by the absorption-averaged beam radius along the crystal

$$w_{p,eff} = \frac{\int_0^L w_p(z)e^{-\alpha_p z}dz}{\int_0^L e^{-\alpha_p z}dz}. \tag{5.48}$$

Using these equations, the behaviour of a longitudinally-diode-pumped solid-state laser can be calculated in a simple way to a good approximation. In order to determine the optimum position of the focus, the minimum of $w_{p,eff}$ with respect to

z_0 has to be found. However, as Eq. (5.48) cannot be calculated analytically, this can only be solved numerically. Therefore, another possibility can be investigated, leading to an analytical solution: As the threshold depends on the square of the pump beam radius, another effective pump beam radius $w'_{p,eff}$ can be defined, the quadratic effective pump beam radius

$$w'_{p,eff} = \sqrt{\frac{\int_0^L w_p^2(z)e^{-\alpha_p z}dz}{\int_0^L e^{-\alpha_p z}dz}}. \qquad (5.49)$$

As there is no root in the integral expression, the integrals can both be calculated analytically, resulting in

$$w'_{p,eff} = \sqrt{r_p^2 + \left[\frac{(z_0\alpha_p - 1)^2 + 1}{\alpha_p^2} - \frac{L^2 - 2Lz_0 + 2\frac{L}{\alpha_p}}{e^{\alpha_p L} - 1}\right]\tan^2\theta_i}. \qquad (5.50)$$

The minimum of this expression with respect to z_0 can be found easily. As the minimum of $w'_{p,eff}$ coincides with the minimum of $w'^2_{p,eff}$, we use

$$\frac{\partial w'^2_{p,eff}}{\partial z_0} = 0 \qquad (5.51)$$

to find the minimum quadratic effective pump spot radius. This yields an optimum focus position of

$$z_{0,opt} = \frac{1}{\alpha_p} - \frac{1 - \eta_{abs}}{\eta_{abs}}L, \qquad (5.52)$$

which approximately minimizes the threshold pump power. Thus, the approximate optimum focus position can be directly calculated from the laser medium parameters. It is surprisingly independent of the divergence angle θ_i and the pump spot diameter r_p.

A second pumping geometry especially suited for high power lasers or amplifier heads is transverse or side pumping, in which the laser diodes are aligned along the side of a laser rod. The pump radiation is then absorbed in a single transverse passage through the rod. This is of course, only possible and efficient for laser media which show a pump absorption length smaller than the rod diameter. This may be achieved using high-spectral-brightness laser diodes pumping the Nd^{3+} ion on its peak absorption, which allows this side-pumping. A sectional view of such a pump arrangement is shown in Fig. 5.13. A big challenge in this layout is the construction of the pump chamber and the complex water flow between the different diode heat sinks and the central flow tube around the laser crystal. As a result of the fact that the diode heat sinks are usually connected electrically in series, they exhibit a potential difference between each other and need to be electrically mutually insulated. As the cooling water, however, forms one closed circuit, one has to control the conductivity of the water, usually below 5 µS, in order to avoid electro corrosion between the different electrical potentials in the cooling circuit.

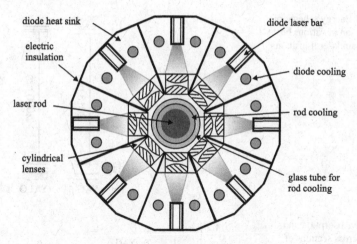

Fig. 5.13 Schematic setup of a transverse diode-pumped rod laser [3]

Recently, optical pumping at around 875 nm has become more and more popular, as this pump line directly excites the ions to the upper laser level. Thus no fast relaxation is necessary and the laser quantum efficiency is increased through reduction of the quantum defect. This results in a lower heat load per pump transition and thus in a higher laser output power, which can be reached before thermal effects become significant.

Applications

The Nd^{3+} laser, especially using the YAG host, is widely used in industry and research applications. In industry, the lasers are mostly used for marking and engraving applications, for spot welding and line welding as well as for hole drilling. In research the laser is mostly used as a high power pump source with high beam quality compared to laser diodes, either by using the laser radiation itself or by frequency doubling of the 1.064 µm line to 532 nm, which is an ideal pump for the Ti:sapphire laser, as discussed in Sect. 5.2.3.

5.2.2 The Tm^{3+}-Laser

The Tm^{3+} laser is a typical quasi-three-level laser operating around 1.9–2 µm. As an example, Fig. 5.14 shows the energy level scheme in three different hosts and Fig. 5.15 the corresponding emission and absorption cross sections for Tm^{3+}:YAG. The specialty of Tm^{3+} lasers is their unique pumping scheme allowing for the use of highly efficient ∼790 nm AlGaAs laser diodes for pumping. As an example, Fig. 5.16 shows the corresponding pump absorption cross section for Tm^{3+}:YAG. This is based on a **cross-relaxation process** $^3H_4 + ^3H_6 \rightarrow 2 \times ^3F_4$, which is very

Fig. 5.14 Energy scheme of
the Tm^{3+} ion in various hosts
and the main laser transitions

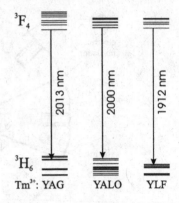

Fig. 5.15 Absorption and
emission cross sections of
Tm^{3+}:YAG around 2 μm

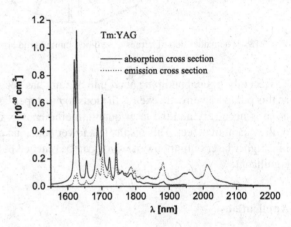

Fig. 5.16 Pump absorption
cross section for Tm^{3+}:YAG

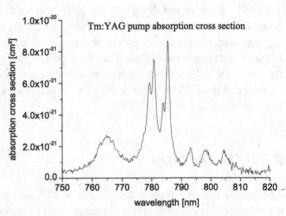

efficient in nearly all hosts. This process is shown in Fig. 5.17 on the example of
a Tm^{3+}:YLF laser: The medium is pumped at a wavelength of 792 nm into the
$^3H_4$ manifold. Owing to the close match in energy between the $^3H_4 - {}^3F_4$ transition
and the $^3H_6 - {}^3F_4$ transition, the excited Tm^{3+} ion makes a transition to the $^3F_4$

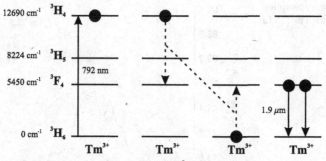

Fig. 5.17 Cross-relaxation process between two Tm^{3+} ions in YLF

manifold. This transition transfers the corresponding energy to a second, unexcited Tm^{3+} ion, which uses this energy to excite its $^3F_4$ manifold. Therefore, two excited ions in the upper laser manifold have been created by one absorbed pump photon. Thus, the number of extracted laser photons per absorbed pump photon, which is called **quantum yield**, can be greater than 100 % in Tm^{3+} lasers.

This process can, of course, also occur in the reverse direction, i.e. that of two Tm^{3+} ions in the $^3F_4$ manifold, from which one can make a transition to the ground-state manifold $^3H_6$, whilst transferring the generated energy to the other ion to excite it from the $^3F_4$ manifold into the $^3H_4$ manifold. This process is called **upconversion**.

Different Hosts for Thulium Lasers

Among the hosts presented in Table 5.1, YLF (YLiF$_4$) provides the highest fluorescence lifetimes and reasonable high pump absorption and laser emission cross-sections, shown in Fig. 5.18 and Fig. 5.19, resulting in low saturation intensities for efficient laser operation. As YLF is a birefringent laser host, different absorption and emission cross sections are needed depending on the polarization of the light with respect to the crystallographic axes. However, the relative re-absorption and thermal lower level population is higher than in YALO or YAG, as shown in Fig. 5.20. YALO, and especially YAG, on the other hand suffer from high saturation intensities. So high brightness pump diode systems are necessary to provide low threshold operation. For the two most important fiber materials, ZBLAN and silica, the spectroscopic data are summarized in Table 5.2. Owing to the very low phonon energy of ZBLAN the fluorescence lifetime is not strongly affected by multi-phonon relaxation. However, in silica the spontaneous lifetime τ_{sp} is about 4.75 ms, which is dramatically reduced by a strong multi-phonon relaxation. Whilst the saturation intensity of Tm^{3+}:ZBLAN is comparable with the ones presented for the crystalline hosts in Table 5.1, Tm^{3+}:silica shows a ten-fold higher saturation intensity compared with ZBLAN. Due to the amorphous nature of the glasses, the optical transitions are inhomogeneously broadened, resulting in broad absorption and emission bands. An example is given for ZBLAN glass in Fig. 5.21 for the laser transition and in Fig. 5.22 for the most important pump band. Also, in the glass media the cross-relaxation pumping is very efficient and results in high overall laser efficiencies.

Fig. 5.18 Pump absorption cross section for Tm^{3+}:YLF

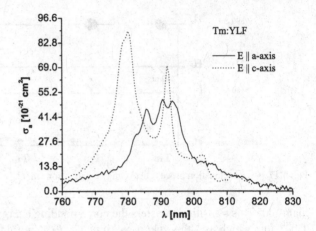

Fig. 5.19 Laser emission cross section for Tm^{3+}:YLF

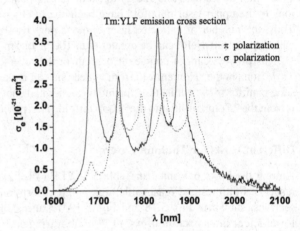

Fig. 5.20 Calculated emission-to-absorption cross-section ratio and lower laser level population of Tm^{3+} in different hosts as a function of the crystal temperature

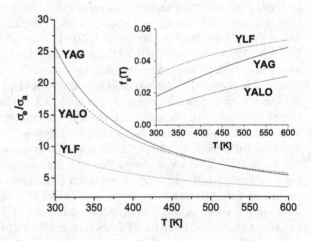

Table 5.1 Important data on Tm^{3+}-doped laser hosts. Some data are taken from [5–9]

Host crystal	YAG	YALO	YLF
$^3F_4$ levels [cm^{-1}]	5556, 5736, 5832, 5901, 6041, 6108, 6170, 6224, 6233	5622, 5627, 5716, 5722, 5819, 5843, 5935, 5965	5605, 5757, 5757, 5760, 5827, 5944, 5967, 5967, 5977
$f_{u,0}$	0.459	0.228	0.286
τ_f [ms]	10.5	5.0	15.6
$^3H_6$ levels [cm^{-1}]	0, 27, 216, 241, 247, 252, 588, 610, 690, 730	0, 3, 65, 144, 210, 237, 271, 282, 313, 440, 574, 628, 628	0, 31, 31, 56, 282, 310, 324, 327, 327, 374, 375, 375, 409
$f_{g,(i)}$	0.018 (6)	0.010 (12)	0.032 (9)
λ_s [nm]	2013	2000	1912
$\sigma_e(\lambda_s)$ [10^{-21} cm^2]	1.53	5.0	4.0 (π)
$\frac{\sigma_e(\lambda_s)}{\sigma_a(\lambda_s)}$	25.8	22.3	9.05
I_{sat}^s [kW/cm^2]	5.91	3.80	1.50
$^3H_4$ levels [cm^{-1}]	12607, 12679, 12747, 12824, 12951, 13072, 13139, 13159	12515, 12574, 12667, 12742, 12783, 12872, 12885, 12910, 12950	12621, 12621, 12644, 12644, 12741, 12825, 12831, 12831, 12831
λ_p [nm]	786	795	792
$\sigma_{a,p}(\lambda_p)$ [10^{-21} cm^2]	8.67	7.5	4.0 (σ), 6.0 (π)
$\frac{\sigma_{e,p}(\lambda_p)}{\sigma_{a,p}(\lambda_p)}$	0.63	1.05	0.88
I_{sat}^p [kW/cm^2]	15.1	9.57	3.62

Table 5.2 Spectroscopic data on Tm^{3+}-doped glasses. Some data are taken from [2, 4]

Host glass	ZBLAN	Silica
τ_f [ms]	10.9	0.34
λ_s [nm]	1940	1970
$\sigma_e(\lambda_s)$ [10^{-21} cm^2]	0.93	2.6
$\frac{\sigma_e(\lambda_s)}{\sigma_a(\lambda_s)}$	23.6	32.3
I_{sat}^s [kW/cm^2]	9.69	110.6
λ_p [nm]	791	790
$\sigma_{a,p}(\lambda_p)$ [10^{-21} cm^2]	3.25	9.93
$\frac{\sigma_{e,p}(\lambda_p)}{\sigma_{a,p}(\lambda_p)}$	1.44	~1
I_{sat}^p [kW/cm^2]	23.8	276
E_p^{max} [cm^{-1}]	590	1100

Fig. 5.21 Absorption and emission cross sections of Tm^{3+}:ZBLAN around 2 μm

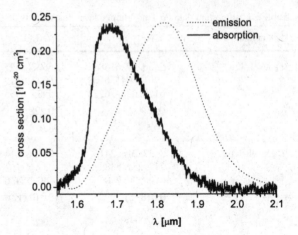

Fig. 5.22 Pump absorption cross section of Tm^{3+}:ZBLAN around 790 nm

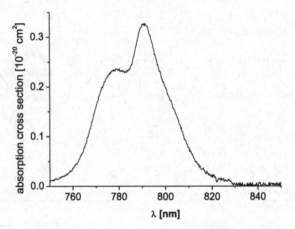

The following section gives an overview over energy-transfer processes and its descriptions.

Energy-Transfer Processes

Fast energy-transfer rates are of the order of 10^7 s^{-1}, whilst the interaction between the active ion and the host phonons occurs at a much faster rate on the order of 10^{11} s^{-1} [10]. Therefore, energy transfer can be seen as an incoherent process and Fermi's golden rule may be applied to the interaction Hamiltonian between the interacting electrons of the donor ion (index D) and the acceptor ion (index A), given by

$$\mathbb{H}_{DA} = \frac{1}{2\kappa} \sum_{i,j} \frac{e^2}{|\vec{r}_i^{D} - \vec{r}_j^{A}|}. \tag{5.53}$$

This corresponds to an interaction that is caused by the electric or magnetic field of the ions, with the electric field contribution being several orders of magnitude

Fig. 5.23 Most important energy-transfer processes in which donor D and acceptor A can be the same or different ion species (denoted by the rare-earth ions RE 1 or RE 2)

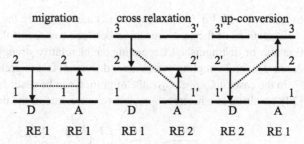

stronger than that of the magnetic field contribution [11]. Here, $\vec{r}$ are the positions of the electrons in ions D and A, κ is related to the polarizability of the medium and the sum is over all electrons in the corresponding ion [14]. Applying Fermi's golden rule and a multipole expansion of the interaction Hamiltonian results in an interaction rate [15, 17]

$$W_{DA} = \frac{C^{dd}}{R^6} + \frac{C^{dq}}{R^8} + \frac{C^{qq}}{R^{10}} + \cdots, \tag{5.54}$$

in which R is the distance between the two ions and the different constants C represent dipole-dipole (dd), dipole-quadrupole (dq) and quadrupole-quadrupole (qq) contributions. As long as the ion spacings R are not too small, the first term in Eq. (5.54) dominates. The interaction lengths may then correspond to several nm [15, 16]. A second interaction mechanism can be caused by the direct overlap between the electronic wave functions of the donor ions and the acceptor ions [18–20]; however, this may only occur at very short inter-ionic distances between the donor and the acceptor, and therefore, only for extremely high dopant concentrations. A special case of this energy-transfer process is the super-exchange, where, the donor and acceptor wave functions overlap not directly, but both overlap with an intermediate ligand ion [21].

For the dipole-dipole interaction, the coupling parameter was shown to be linked to the overlap between the donor emission cross section $\sigma_e^D(\lambda)$ and the acceptor absorption cross section $\sigma_a^A(\lambda)$ by Dexter [17]

$$C_{DA}^{dd} = \frac{9\chi^2 c}{16\pi^4 n^2} \int \sigma_e^D(\lambda) \sigma_a^A(\lambda) d\lambda. \tag{5.55}$$

Therein, $\chi^2 \sim \frac{2}{3}$ accounts for an orientational average.

Whilst this approach deals with the microscopic interaction between two ions, it is important to have a description of the macroscopic behaviour of an active medium in order to model its excitation processes. The most important energy-transfer processes occurring in rare earth doped solid-state materials are shown in Fig. 5.23. They are migration, cross relaxation and upconversion. Migration is the energy transfer between ions of the same species involving the same transition levels in both ions, called donor-donor transfer. Thus, the corresponding coupling parameter is given by

$$C_{DD} = \frac{3c}{8\pi^4 n^2} \int \sigma_e(\lambda) \sigma_a(\lambda) d\lambda \tag{5.56}$$

by analogy with Eq. (5.55). Here, $\sigma_e(\lambda)$ and $\sigma_a(\lambda)$ are the absorption and emission cross sections of the same transition between two manifolds. Three types of migration may be distinguished, depending on the relative strength of the donor-donor and donor-acceptor coupling parameters: diffusion, fast migration and super-migration.

In the case of $C_{DD} \ll C_{DA}$, the migration can be described by a diffusion process [22], resulting in a macroscopic de-excitation rate of the donors of

$$\frac{\partial N_D}{\partial t} = -\frac{16\pi^2}{3}\left(\frac{1}{2}\right)^{\frac{3}{4}} C_{DA}^{\frac{1}{4}} C_{DD}^{\frac{3}{4}} N_D^0 N_D N_A, \qquad (5.57)$$

where N_D^0 is the total density of donors that can contribute to the migration process, N_D the density of excited donors and N_A the density of acceptors. This macroscopic rate is valid for long times given by [12]

$$t > \frac{16\pi^3}{9}\frac{C_{DA}}{W_{DA}^2} N_A^2 \approx \frac{16\pi^3}{9}\frac{R_{DA}^{12}}{C_{DA}} N_A^2, \qquad (5.58)$$

with R_{DA} being the donor-acceptor distance.

The fast migration ($C_{DD} \gg C_{DA}$) is often the dominant process with donor-donor coupling parameters C_{DD} that are some orders of magnitude larger than those for the donor-acceptor energy transfer C_{DA} given by Eq. (5.55). This can be explained as the donor-acceptor interaction depends on two different transitions, which need to overlap spectrally to yield a large energy-transfer parameter, whilst the donor-donor interaction relies upon the same transition, so the overlap is guaranteed by Eq. (2.32). In the case of fast migration, a donor excitation can migrate significantly before it interacts with an acceptor, thus increasing the macroscopic probability for the donor-acceptor transfer processes to occur. Whenever $C_{DD} \geq C_{DA}$, the macroscopic de-excitation rate can be described by the hopping model [23], resulting in

$$\frac{\partial N_D}{\partial t} = -\pi\left(\frac{2\pi}{3}\right)^{\frac{5}{2}} \sqrt{C_{DA}C_{DD}} N_D^0 N_D N_A \qquad (5.59)$$

for long times as given by Eq. (5.58). In the case of upconversion ($N_D^0 = N_{RE2}$, $N_D = N_{2'}$ and $N_A = N_2$) this rate equation is often rewritten as

$$\frac{\partial N_{2'}}{\partial t} = -k_{up} N_2 N_{2'} \qquad (5.60)$$

with

$$k_{up} = \pi\left(\frac{2\pi}{3}\right)^{\frac{5}{2}} \sqrt{C_{DA}^{up} C_{DD}^{up}} N_{RE2}, \qquad (5.61)$$

$$C_{DA}^{up} = \int \sigma_{2'\to1',e}(\lambda)\sigma_{2\to3,a}(\lambda)d\lambda, \qquad (5.62)$$

$$C_{DD}^{up} = \int \sigma_{2'\to1',e}(\lambda)\sigma_{1'\to2',a}(\lambda)d\lambda, \qquad (5.63)$$

whilst for cross relaxation ($N_D^0 = N_{RE1}$, $N_D = N_3$ and $N_A = N_{1'}$) one often finds that

$$\frac{\partial N_3}{\partial t} = -k_{cr}N_{1'}N_3 \tag{5.64}$$

with

$$k_{cr} = \pi \left(\frac{2\pi}{3}\right)^{\frac{5}{2}} \sqrt{C_{DA}^{cr} C_{DD}^{cr}} N_{RE1}, \tag{5.65}$$

$$C_{DA}^{cr} = \int \sigma_{3\to 2,e}(\lambda)\sigma_{1'\to 2',a}(\lambda)d\lambda, \tag{5.66}$$

$$C_{DD}^{cr} = \int \sigma_{3\to 1,e}(\lambda)\sigma_{1\to 3,a}(\lambda)d\lambda. \tag{5.67}$$

Upconversion and cross relaxation are reverse processes on the microscopic scale, which are thermodynamically linked to each other. Using Eqs. (2.32), (5.55) the relation

$$\frac{C_{DA}^{cr}}{C_{DA}^{up}} = \frac{\int \sigma_{3\to 2,e}(\lambda)\sigma_{1'\to 2',a}(\lambda)d\lambda}{\int \sigma_{2'\to 1',e}(\lambda)\sigma_{2\to 3,a}(\lambda)d\lambda} = \frac{Z_{2'}Z_2}{Z_{1'}Z_3} \tag{5.68}$$

can be deduced as the ratio of the microscopic transfer parameters. It depends only on the partition functions Z_i of the involved manifolds. In the special, but most important case, in which RE 1 and RE 2 are the same ion species, this expression simplifies to

$$\frac{C_{DA}^{cr}}{C_{DA}^{up}} = \frac{\int \sigma_{3\to 2,e}(\lambda)\sigma_{1\to 2,a}(\lambda)d\lambda}{\int \sigma_{2\to 1,e}(\lambda)\sigma_{2\to 3,a}(\lambda)d\lambda} = \frac{Z_2^2}{Z_1 Z_3} \tag{5.69}$$

As a result of the principal difference between the migration processes in upconversion and cross relaxation, such a simple relation cannot be set up for the rate-equation parameters k_{cr} and k_{up}. However, it can be shown from the equations above that

$$\frac{k_{cr}}{k_{up}} = \frac{N_{RE1}}{N_{RE2}} \frac{Z_{2'}}{Z_{1'}Z_3} \sqrt{Z_1 Z_2 \frac{\int e^{-\frac{hc}{k_B T\lambda}}\sigma_{1\to 3,a}^2(\lambda)d\lambda}{\int e^{-\frac{hc}{k_B T\lambda}}\sigma_{1'\to 2',a}^2(\lambda)d\lambda}}, \tag{5.70}$$

and for the case of identical species this simplifies to

$$\frac{k_{cr}}{k_{up}} = \frac{Z_2}{Z_3} \sqrt{\frac{Z_2}{Z_1} \frac{\int e^{-\frac{hc}{k_B T\lambda}}\sigma_{1\to 3,a}^2(\lambda)d\lambda}{\int e^{-\frac{hc}{k_B T\lambda}}\sigma_{1\to 2,a}^2(\lambda)d\lambda}}. \tag{5.71}$$

The regime of super-migration is reached when the actual acceptor concentration c_A, i.e. the probability of occupation of a possible lattice site by an acceptor, exceeds the critical concentration c^*,

$$c_A > c^* = \left(\frac{C_{DA}}{C_{DD}}\right)^{\frac{1}{8}}. \tag{5.72}$$

Fig. 5.24 Atomic structure
of the Al_2O_3 host crystal [28]

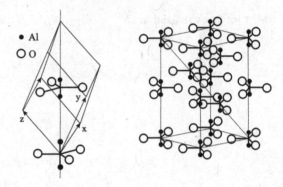

Owing to $c_A < 1$ the super-migration can only occur for $C_{DD} > C_{DA}$. Then the fast migration distributes the energy over all donors before the donor-acceptor transfer can occur.

It has to be noted that even if the absorption and emission spectra in Eq. (5.55) do not overlap, an energy-transfer process may still exist for this transition; provided the excess energy that would be needed to make the cross sections overlap is compensated for by the emission or absorption of a lattice phonon [24–27].

5.2.3   The Ti^{3+}:Al_2O_3 Laser

In contrast to the two solid-state lasers discussed before, which are examples of rare-earth-ion-based laser media, the Ti^{3+}:sapphire laser (Ti^{3+}:Al_2O_3) uses the trivalent transition-metal titanium ion as the active ion. The host medium is single-crystalline aluminum oxide Al_2O_3, the same host as used in the first laser, the ruby laser (Cr^{3+}:Al_2O_3). In this context, it should be noted that the name "Ti^{3+}:sapphire laser" is tautological, as "sapphire" is already the name for Ti^{3+}-doped aluminum oxide, just as "ruby" denotes the Cr^{3+}-doped aluminum oxide. Thus, one could also speak of the "sapphire laser" itself. In the following, we will discuss the special energy structure and the resulting laser properties arising from the transition-metal nature of the active ion, as well as some applications of the Ti^{3+}:Al_2O_3 laser.

The Laser Medium

Being a $3d$ transition metal, the Ti^{3+} ion exhibits a single $3d$ electron in its outer shell, which is the "active laser" electron. The free ion is thus similar to a hydrogen atom and higher-lying electronic states cannot be excited by the energies accessible during optical pumping. The whole pump and laser process thus takes place between the different levels that are created by the splitting of the five-fold-degenerate free-ion ground state ^2D in the host matrix. This host matrix is shown in Fig. 5.24.

In contrast to the splitting scheme discussed in Fig. 2.3 for the rare-earth ions, the optically active electron here is not shielded from the crystal field. This causes

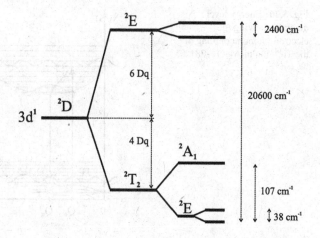

Fig. 5.25 Energy level diagram of the crystal-field splitting of the Ti^{3+} ion in Al_2O_3 [28]

a crystal-field splitting which dominates the spin-orbit splitting. Whilst the actual energy of this splitting is, of course, a host- and ion-dependent value, the theoretical description of the strong-crystal-field splitting can be generalized. In this perturbational description, the scale of the splitting energy is given by the parameter product Dq, in which

$$D = \frac{1}{4\pi\epsilon_0} \frac{35}{4} \frac{Ze^2}{a^5} \qquad (5.73)$$

accounts for the strength of the crystal field, caused by the charge of the ligand $-Ze$ at a distance a from the central ion, and

$$q = \frac{2}{105} \langle 3d|r^4|3d \rangle \qquad (5.74)$$

is proportional to the quantum-mechanical radial integral of the $3d$ wave functions that has to be calculated to obtain the energy difference. The ground-state splits by a total amount of $10Dq$, for which the ratio between the amount of energy decrease of the lower state and the amount of increase of the upper state depends on the symmetry of the external ligand ions and the resulting degeneracy of the final levels. In the case presented here, there are the six oxygen ions around the Ti^{3+} ion, resulting in an octahedral coordination, causing a splitting into a doubly-degenerate ^2E state and a triply-degenerate $^2T_2$-state. As the total energy change of the splitting needs to be zero, the ^2E state is raised $6Dq$ above the initial, free-ion ground state, whilst the $^2T_2$-state is lowered by $4Dq$ with respect to the initial ground state as shown in Fig. 5.25. As these octahedra are trigonally distorted, as can be seen in Fig. 5.24, the $^2T_2$-state is split further. The ^2E excited state degeneracy is also lifted by the so called **Jahn-Teller effect**, which states that the degeneracy of an electronic state in a non-linear complex will be lifted by a spontaneous deformation of the surrounding lattice. However, strictly speaking, no optical dipole transitions should occur between these states as they all have even parity. The fact that there are optical dipole transitions is a direct result of the breaking of inversion symmetry as soon as an Al^{3+} ion is replaced by a Ti^{3+} ion. This causes a mixing of the odd-parity wave

Fig. 5.26 Energy level diagram showing the electron-phonon coupling of the Ti^{3+} ion in Al_2O_3 [28]

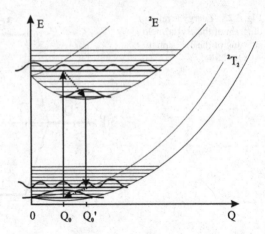

functions of the oxygen ions with those on the Ti^{3+} ion, allowing optical-dipole transitions within the split ground state.

Up to this point, there is no principle difference of this energy scheme with the ones discussed for the rare-earth ions. However, the energy levels here are a direct consequence of the crystal field, and thus, of the spatial positions of the oxygen ions in the crystal lattice. Their energies are therefore very sensitive to this configuration. The phonons of the host crystal now cause vibrations of the oxygen ions and thus a modulation of the crystal field, which will have an influence on to the energies of those levels. Quantum-mechanically, this strong electron-phonon coupling results in so-called **vibronic states**, mixed states between the electronic states of the ion and the phonon states of the lattice. They can be described by the **configurational coordinate model**. In this model, the energy of the levels is plotted as a function of the configurational coordinate Q, which can be seen as a parameter describing the distance of the oxygen atoms of the vibrating octahedra with respect to the Ti^{3+} ion. In this description, the energy variation of a level can be described for small changes in configuration by a parabolic potential as shown in Fig. 5.26, and thus at each level, a harmonic oscillator can be assumed. For the ground state, this results in three paraboloids oriented around the origin $Q = 0$, from which Fig. 5.26 shows a cut along one radial Q-axis through the minimum of one of the paraboloids. For the excited state, two paraboloids result, which are also shifted outward from the origin $Q = 0$. As the Jahn-Teller effect is different for the ground and the excited state, it causes a different lattice distortion, and thus, the corresponding configurational coordinate of the parabola minimum Q_0 and Q_0' is also different. Taking into account that the transition probability depends on the overlap of the wave functions, which are also indicated in Fig. 5.26, a strong wavelength shift exists between absorption and emission. This is called **Franck-Condon shift**. After the absorption process, the excitation will quickly thermalize within the upper harmonic oscillator levels owing to the strong interaction with the phonons, which created these levels. Thus, the emission process with the highest probability will start from the lowest levels in the excited state. However, from this state only transitions to higher states

Fig. 5.27 Upper-state
fluorescence lifetime of the
Ti^{3+} ion in Al_2O_3 as a
function of the temperature
[28]

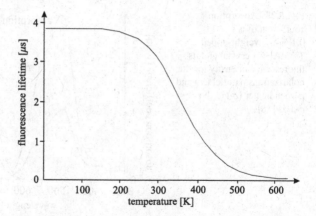

within the ground-state parabola yield a high overlap in wavefunction, as the coordinates of the minima are different (**Franck-Condon principle**). This causes the
strong wavelength shift between absorption and emission.

A second effect of the vibronic states is the large fluorescence bandwidth, and
thus, the tunability of the $Ti^{3+}:Al_2O_3$ laser. Owing to the spatially oriented three
paraboloids in the ground state, a transition from an excitation at a fixed coordinate
Q can occur towards a huge number of possible final states and phonon energies in
the ground state.

The $Ti^{3+}:Al_2O_3$ laser crystals are grown by the Czochralski method. In contrast
to $Nd^{3+}:YAG$, where the dopant concentration is limited by the ionic radius of the
neodymium ion to approximately 1 %, or compared with $Tm^{3+}:YAG$, in which the
thulium ion concentration can reach several 10 %; the Al_2O_3 crystal can only be
doped with Ti^{3+} on the order of 0.1–0.25 %. One reason for this low concentration
is the higher ionic radius of Ti^{3+} (0.067 nm) compared with Al^{3+} (0.053 nm), the
other reason can be related to the Jahn-Teller effect, as a higher doping concentration
would cause a ground-state degeneracy, which itself would result in a Jahn-Teller
splitting, creating a lattice deformation that is not compatible with the Al_2O_3 lattice.

Laser Parameters

The lifetime of the $Ti^{3+}:Al_2O_3$ fluorescence at room temperature is around 3.1 µs,
and thus three orders of magnitude shorter than the fluorescence lifetimes of many
rare-earth doped crystals, such as those using Tm^{3+}, Ho^{3+} or Er^{3+} as the active
ion. Owing to the electron-phonon coupling, this lifetime strongly depends on the
crystal temperature as shown in Fig. 5.27. As the optical transitions are a result of
the statically broken inversion symmetry, mixing odd-parity wave functions into the
ground state, and not a result of the vibronic states, the spontaneous optical emission
rate τ_{sp}^{-1} is temperature independent. However, also non-radiative transitions from
the excited state into the ground state exist. These relaxation transitions show a rate
τ_r^{-1}, resulting in a total fluorescence lifetime of

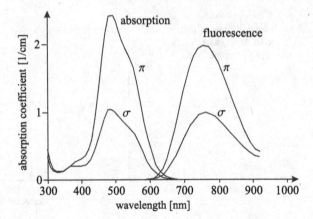

Fig. 5.28 Absorption coefficient of a 0.1 %-by-weight-doped Ti^{3+}:Al_2O_3 crystal and its fluorescence intensity for the polarizations parallel (π) and perpendicular (σ) to the c-axis [28]

$$\tau_{tot}^{-1} = \tau_{sp}^{-1} + \tau_r^{-1}. \tag{5.75}$$

This relaxation rate is strongly temperature dependent and is caused by a tunneling of the upper level excitations in the ^2E parabolas into the ground-state parabolas. As the energy gap (tunnel distance) is smaller for highly excited levels, the increasing fractional excitation of the higher levels in the ^2E parabolas with temperature causes a strong increase in the non-radiative transition rate. As can be seen from Fig. 5.27, this process starts at a temperature around 200 K and then quickly reduces the fluorescence lifetime for higher temperatures. The quantum yield of the laser at room temperature, which is the amount of fluorescence photons with respect to the total amount of transitions, thus results in

$$\eta_{QY} = \frac{\tau_{tot}(300\ K)}{\tau_{sp}} = \frac{3.1\ \mu s}{3.85\ \mu s} = 0.8. \tag{5.76}$$

The fluorescence of Ti^{3+}:Al_2O_3 is a maximum for light polarized parallel to the crystallographic c-axis as shown in Fig. 5.28 together with the absorption coefficient. This results in a peak emission cross section for this polarisation of 3.5×10^{-19} cm^2 at 795 nm, which has been determined from its fluorescence spectrum using the Füchtbauer-Ladenburg relation Eq. (1.77) in Fig. 5.29. It can be clearly seen that the emission cross section is shifted towards lower wavelengths with respect to the fluorescence. The absorption band is very broad, resulting in a large variety of possible pump sources. Whilst in the past mainly Ar^+ ion lasers at 514 nm have been used as a pump, today, frequency-doubled Nd^{3+}:YAG, Nd^{3+}:YVO$_4$ or Nd^{3+}:YLF lasers are used owing to their much higher efficiency compared with the Ar^+ laser. Also pulsed pumping by flashlamps is possible, however, as a result of the short upper level lifetime, high pump intensities and short pump pulses are necessary, which needs specially designed low-inductance flashlamp circuits and flashlamps for high operation lifetime. Especially the flashlamp design is critical when it comes to high pump energies of several 100 J as the short pulse width on the order of 2–10 μs also corresponds to a low flashlamp explosion energy, see Eq. (5.8).

Fig. 5.29 Emission cross
section and fluorescence
intensity of a Ti^{3+}:Al_2O_3
crystal for the polarization
parallel (π) to the c-axis [28]

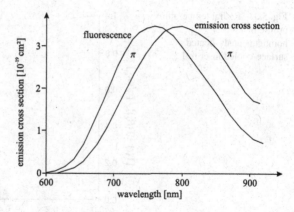

Fig. 5.30 Absorption and
emission cross sections of
Cr^{2+}:ZnSe [29]

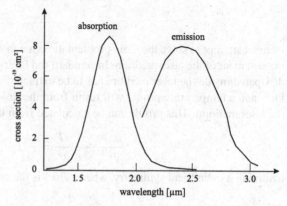

Applications

The main application of the Ti^{3+}:Al_2O_3 laser today arises from its large tuning and
amplification range. It is the generation of ultra-short pulses, already discussed in
Sect. 4.2, as well as the amplification of ultra-short pulses discussed in Sect. 4.2.3,
where these lasers are dominant. Another important application is in spectroscopy,
to generate broadly-tunable radiation with a small linewidth. Similar spectral properties can be achieved with Cr^{3+}:$LiSrAlF_6$ and Cr^{3+}:$LiCaAlF_6$, which show an
emission band shifted towards the infrared compared with Ti^{3+}:Al_2O_3. However,
the fluoride crystals are hygroscopic, which is a drawback for efficient cooling of
the laser rods. A comparable laser medium in the 2.3 μm region is Cr^{2+}:ZnSe, which
shows a broad emission range at 2–3 μm, as shown in Fig. 5.30. This allows the generation and amplification of ultra-short pulses in the mid-infrared spectrum.

5.3   Special Realisations of Lasers

In the following section we will investigate two of the most important laser geometries today, which are the fiber laser and the thin-disk laser. Both of these laser archi-

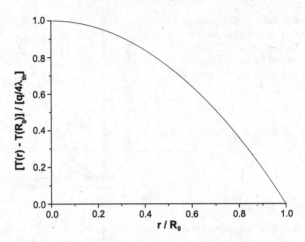

Fig. 5.31 Radial temperature profile in a homogeneously-heated, surface-cooled laser rod

tectures attempt to solve the main problematic issue in solid-state lasers: the thermal management of the laser medium. In standard rod lasers as discussed above, the heat dissipated inside the laser medium has to be extracted through the outer crystal surface, and a temperature profile will result from the finite heat transfer coefficient of the laser medium. This profile can be calculated from the heat-transfer equation [3]

$$\frac{\partial^2 T}{\partial r^2} + \frac{1}{r}\frac{\partial T}{\partial r} = \frac{q}{\lambda_{th}}, \tag{5.77}$$

assuming a cylindrical symmetry, where, $T(r)$ is the rod temperature,

$$q = \frac{P_{therm}}{\pi R_0^2 L} \tag{5.78}$$

the volumetric heat load, L the length of the rod, P_{therm} the power dissipated as heat in the medium and λ_{th} the heat-transfer coefficient of the medium. In the case of a homogeneous heat load q, this results in a parabolic temperature profile

$$T(r) = T(R_0) + \frac{q}{4\lambda_{th}}\left(R_0^2 - r^2\right), \tag{5.79}$$

with R_0 being the radius of the rod and $T(R)$ the temperature on the outer crystal surface. This temperature profile is shown in Fig. 5.31 and has two important influences on the laser medium, which are discussed in the following section.

5.3.1 Thermal Lensing and Thermal Stress

Thermal Lensing

As the index of refraction of the laser medium is usually temperature dependent, this temperature profile with the gain medium will create a refractive index distribution resulting in the formation of a **thermal lens**. The corresponding index distribution is given by

$$n(r) = n_0 + \frac{\partial n}{\partial T}(T(r) - T(R_0)), \tag{5.80}$$

wherein $n_0 = n(R_0)$ is the refractive index of the rod at the surface. The thermal lens can have a positive or negative focal length depending on the sign of the thermal index coefficient $\frac{\partial n}{\partial T}$ of the laser medium. YAG, for example, has $\frac{\partial n}{\partial T} = 9.9 \times 10^{-6}$ K^{-1} and thus a positive thermal lens, whilst, for example YLF, with $\frac{\partial n}{\partial T} = -2 \times 10^{-6}$ K^{-1} for a polarization along the a-axis, shows a negative, i.e. diverging, thermal lens. Taking into account that a parabolic index profile

$$n(r) = n_0 - \frac{1}{2}n_2 r^2, \tag{5.81}$$

which is constant in axial direction along a (rod) length L, acts like a lens with a focal length of

$$f = \frac{1}{n_2 L}, \tag{5.82}$$

we obtain for the thermal lens

$$f_{th} = \frac{2\lambda_{th}\pi R_0^2}{\frac{\partial n}{\partial T} P_{therm}}, \tag{5.83}$$

where, it has to be taken into account that Eq. (5.82) is only valid for $f \gg L$. In some laser media, such as ZnSe or YAlO$_3$ (YALO), the thermal lensing can be strong enough that focal lengths shorter than the rod may result, especially for long crystals. But this is usually connected simultaneously to a very strong beam quality degradation as a result of aberrations caused by the non-parabolic temperature profile, and is therefore avoided in most lasers.

However, the measured values of thermal lenses differ from the simple relation in Eq. (5.83), because additional effects have to be considered: We also have to take into account that the rod will show a local thermal expansion, which will create a thermal stress. This stress itself causes an additional change in the refractive index by the **photoelastic effect**. All these material dependent effects can, however, be summed into one parameter ξ. For isotropic laser media such as YAG, the focal length of the total thermal lens can be expressed by

$$f_{th} = \frac{\pi R_0^2}{\xi P_{therm}}. \tag{5.84}$$

For YAG, a value of $\xi = 5.09 \times 10^{-7}$ $\frac{m}{W}$ has been measured.

Finally, we have to take into account that the total thermal expansion will cause a bulging of the end faces of the rod, which itself is a positive contribution to the total thermal lens. This effect depends on the length of the crystal and results in a modified parameter [3]

$$\xi' = \xi + \frac{\alpha_{th} R_0 (n_0 - 1)}{\lambda_{th} L}, \tag{5.85}$$

which replaces the ξ in Eq. (5.84), where, α_{th} is the coefficient of thermal expansion of the medium.

In designing a laser resonator, the formation of a thermal lens has to be taken into account in order to maintain performance over an extended period. However, from the power dependence of the thermal lens it directly follows that in a laser with a medium that is affected by strong thermal lensing, the resonator has to be calculated and optimized for the operation point of the laser, i.e. for a certain thermal lens that will build up at the nominal pump and output power of the laser. A direct result of thermal lensing will mean that the laser is designed for a certain pump power only. The use of differing pump powers will cause a change in the laser mode size and waist position inside the resonator, and therefore, will lead to a varying overlap with the pump beam. Therefore, the laser slope efficiency can depend on the pump power. Also the true laser threshold can be different from the theoretical one, as with thermal lensing the threshold depends also on the stability range of the cavity, which becomes pump-power dependent.

Thermal Stress

In an actively cooled laser rod the inner volume has a higher temperature, and therefore, a larger thermal expansion than the outer part of the rod. This results in the formation of mechanical stresses. In a cylindrical laser rod, the stresses in radial, tangential and axial (z-axis) direction can be calculated from the temperature distribution in the plain strain approximation, which is valid for long, surface cooled laser media. This results in [49]

$$\sigma_r(z,r) = \frac{\alpha_T E}{1-\nu}\left(\frac{1}{R^2}\int_0^{R_0} T(z,r')r'dr' - \frac{1}{r^2}\int_0^r T(z,r')r'dr'\right)$$

$$\sigma_t(z,r) = \frac{\alpha_T E}{1-\nu}\left(\frac{1}{R^2}\int_0^{R_0} T(z,r')r'dr' + \frac{1}{r^2}\int_0^r T(z,r')r'dr' - T(r,z)\right) \quad (5.86)$$

$$\sigma_z(z,r) = \sigma_r(z,r) + \sigma_t(z,r) = \frac{\alpha_T E}{1-\nu}\left(\frac{2}{R^2}\int_0^{R_0} T(z,r')r'dr' - T(z,r)\right),$$

with σ_r, σ_t and σ_z being the radial, tangential and axial stress components, respectively, E being Young's modulus, ν Poisson's ratio, α_{th} the thermal expansion coefficient and R the radius of the rod. The equation for σ_z is valid for a laser medium where the ends are free to move. Inserting the temperature distribution from Eq. (5.79) for the homogeneously heated laser rod, we obtain

$$\sigma_r = \frac{\alpha_T E}{16\lambda_{th}(1-\nu)}q\left(r^2 - R_0^2\right) = \sigma_0\left(r^2 - R_0^2\right), \quad (5.87)$$

$$\sigma_t = \frac{\alpha_T E}{16\lambda_{th}(1-\nu)}q\left(3r^2 - R_0^2\right) = \sigma_0\left(3r^2 - R_0^2\right), \quad (5.88)$$

$$\sigma_z = \frac{\alpha_T E}{8\lambda_{th}(1-\nu)}q\left(2r^2 - R_0^2\right) = 2\sigma_0\left(2r^2 - R_0^2\right). \quad (5.89)$$

Fig. 5.32 Stress components in a homogeneously-heated, surface-cooled laser rod

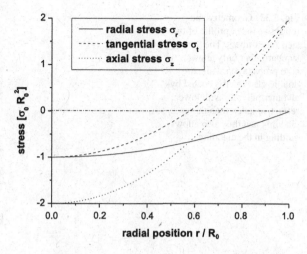

Table 5.3 Thermal-shock parameter of several laser media [3]

Host crystal	YAG	GSAG	Al$_2$O$_3$	SiO$_2$ glass
R_s	$7.9\,\frac{W}{cm}$	$6.5\,\frac{W}{cm}$	$100\,\frac{W}{cm}$	$1\,\frac{W}{cm}$

The stress distributions are plotted in Fig. 5.32. A positive value corresponds to a tensile stress in the corresponding direction, whilst a negative value denotes a compressive stress. These stresses will cause refractive index changes, which is called **stress-induced birefringence**. This can change the polarization of the laser mode and thus degrades polarization quality in laser systems. However, these effects can be compensated for by more complex laser designs, in which several laser rods and polarization rotators are used.

The main problem for high average power lasers, however, is the tensile stress on the outer crystal surface. Owing to the perpendicularity of the tangential and axial stress, the total stress on the surface results in

$$\sigma_{tot} = \sqrt{\sigma_t^2 + \sigma_z^2} = 2\sqrt{2}\sigma_0 R_0^2 \tag{5.90}$$

If this tensile stress exceeds a certain value σ_{max}, it will lead to a growth of microscopic cracks on the outer crystal surface, which finally results in a total crystal fracture. Using Eq. (5.78), this maximum stress corresponds to a maximum power dissipation per crystal length of [3]

$$\frac{P_{therm}^{max}}{L} = 8\pi \frac{\lambda_{th}(1-\nu)}{\sqrt{2}\alpha_{th}E}\sigma_{max} = 8\pi R_s, \tag{5.91}$$

which is independent of the rod diameter, where, R_s is the **thermal-shock parameter**, which is shown in Table 5.3 for several laser hosts. This effect yields for YAG, that a thermal extraction of approximately 200 W/cm will fracture the rod. However, this value depends strongly on the surface finish of the laser rod and the real value at which fracture occurs can differ from this value by up to a factor of three.

Fig. 5.33 Geometry and
refractive index profile of two
step-index fibers. The
standard fiber only allows
core propagation while the
double-clad fiber is coated by
an outer polymer with lower
refractive index than the
cladding and thus also allows
guiding in the cladding

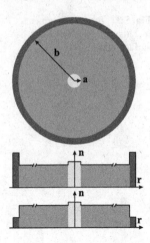

In order to avoid all these temperature-dependent effects to a great extent, the
following laser types have been developed, namely, the fiber laser and the thin-disk
laser.

5.3.2 The Fiber Laser

An optical fiber consists of a core with radius a and a refractive index n_{core}, and a
cladding with radius b, showing a lower refractive index $n_{cladding} < n_{core}$ as shown
in Fig. 5.33. In a fiber laser, this core is in addition doped with laser-active ions,
which are pumped by a pump radiation also guided in the fiber.

Owing to the high surface-to-volume ratio

$$\frac{2\pi a L}{\pi a^2 L} = \frac{2}{a} \tag{5.92}$$

and the small fiber radii $a < 1$ mm, the heat transfer from the active core to the large
surface occurs over a small distance. Thus, the resulting temperature differences are
low, even for the lower thermal conductivities of glass materials. This causes a lower
temperature in the active region, and therefore, a higher laser efficiency, especially
for quasi-three-level lasers. In a rotationally symmetric fiber a parabolic temperature
profile will develop; however, this will only cause a refractive index difference of

$$\Delta n_{therm} \approx \frac{P_{therm}}{4\pi \lambda_{th} L} \frac{\partial n}{\partial T} = 8 \cdot 10^{-6} \tag{5.93}$$

for a $P_{th} = 1$ kW dissipating, $L = 100$ m-long silica fiber ($n \approx 1.45$) with NA $=$
0.04, $\lambda_{th} \approx 1 \frac{W}{K\,m}$ and $\frac{\partial n}{\partial T} = 10^{-5}$ K^{-1}. Comparing this with the refractive-index
difference causing the guiding in the fiber, we obtain

$$\Delta n_{guide} \approx \frac{NA^2_{core}}{2n^2_{core}} = 5 \cdot 10^{-4}. \tag{5.94}$$

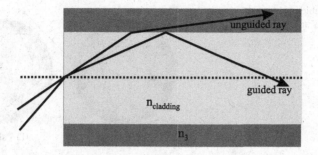

Fig. 5.34   Path of a light ray coupled into the cladding of a fiber

Thus, even for very high thermal power dissipations, the fundamental guiding properties of the fiber stay unchanged, and therefore, the beam quality of the fiber laser will be preserved in the case of a single-mode fiber even for high output powers in the kW range.

Double-Clad Fibers

As a protection the fiber is usually coated by an outer polymer. Its refractive index n_3 therefore determines, whether light may be guided in the cladding or not. As the refractive index profiles consist of several steps, this fiber type is called **step-index fiber**. The cladding diameter is usually much larger than the wavelength, and therefore the propagation within the cladding can be calculated in the scope of geometrical optics. Thus, all light that will hit the fiber end within an acceptance solid angle $\Delta\Omega_0$ with a half angle θ_i will be guided inside the cladding, see Fig. 5.34. This angle can be calculated from the total-internal-reflection angle inside the fiber as

$$\Delta\Omega_0 = 2\pi(1 - \cos\theta_i) = 2\pi\left(1 - \sqrt{1 - \mathrm{NA}_{cladding}{}^2}\right). \tag{5.95}$$

This acceptance cone thus only depends on the numerical aperture of the cladding, given by $\mathrm{NA}_{cladding} = \sqrt{n_{cladding}^2 - n_3^2}$.

In a standard fiber, the polymer usually has a higher refractive index than the cladding, so that no total-internal reflection is possible. This is mostly used for passive fibers, e.g. in telecommunication. In the opposite situation, the cladding can also transport light, and the fiber is called a **double-clad fiber**.

The guiding in the core can be explained in the same way, however, as a result of the smaller core sizes the guiding needs to be calculated by wave optics, which is discussed later. As the guiding core has a higher refractive index than the cladding, the light guided in the cladding can also pass the core of the fiber. This is important for high-power fiber lasers, in which the pump radiation is coupled into the cladding of the fiber, from which it will be absorbed each time it passes the core. The fluorescence is then emitted from the excited ions into the core and the cladding. However,

Fig. 5.35 Double-clad fiber
with symmetric and D-shaped
geometry

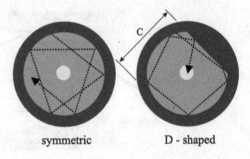

symmetric   D - shaped

only those photons will be amplified to a great extent which only propagate inside
the core. Thus, the laser radiation will be guided in the core. Owing to the smaller
core size, the laser radiation exhibits a much higher beam quality than the pump
radiation. Therefore, the double-clad fiber laser can also be seen as a brightness
transformer (or brightness enhancer), transforming the light of the cladding (with a
low brightness, a high number of modes and therefore a low beam quality) into the
light guided in the core, which shows a high brightness and is emitted into only a
few or a single mode, thus having a very high beam quality. Of-course, this is not a
true transformer, as the emitted light shows a frequency different from the incoming
pump radiation.

However, in double-clad fibers it has to be taken into account that skew waves
can build up in completely rotational symmetric index profiles. These skew waves
do not pass through the core during their propagation inside the fiber, and therefore,
would not be absorbed by the active ions. To obtain a maximum pumping efficiency,
these skew waves have to be suppressed, which can be obtained by breaking the
rotational symmetry of the fiber. This can be done in several ways, from which
the easiest one consists of milling a flat surface on to the fiber preform prior to
fiber pulling. This results in the geometry shown in Fig. 5.35. Usually, about 10–
15 % of the fiber diameter are cut away ($c \approx 1.8b$), creating a linear reflection edge
which breaks the rotational symmetry and makes self-consistent rays that do not hit
the core impossible. Assuming a homogeneous filling of the cladding by the pump
radiation, which is a direct consequence of the large amount of excited modes in the
cladding, the effective pump absorption inside a double-clad fiber can be expressed
by

$$\alpha_{eff} = \frac{A_{core}}{A_{cladding}} \alpha_p = \frac{A_{core}}{A_{cladding}} \sigma_a(\lambda_p) N_1. \tag{5.96}$$

Thus, in a double-clad fiber laser, the pump absorption length can be chosen independently from the length of the laser medium within certain limits. As the cladding
can be fabricated with a high numerical aperture using a low-index outer polymer, a
high numerical aperture of the cladding can be chosen without changing the mode
properties of the fiber core. Thus, very high pump powers can be used. In particular,
fiber coupled laser diodes can directly pump such a fiber laser without the need for
a high-beam-quality pump laser.

Propagation in the Core

The step in the refractive index between core and cladding, together with the diameter of the core, determine how the light propagates in the core. As the core radius a is usually comparable with the wavelength of the laser emission, we have to solve the wave equation, resulting in the existence of modes in a similar way to those in the laser resonators. Here, we use the scalar wave equation in cylindrical coordinates

$$\frac{\partial^2 \Psi}{\partial r^2} + \frac{1}{r}\frac{\partial \Psi}{\partial r} + \frac{1}{r^2}\frac{\partial^2 \Psi}{\partial \phi^2} + \frac{\partial^2 \Psi}{\partial z^2} + n(r)^2 k_0^2 \Psi = 0, \tag{5.97}$$

where, Ψ corresponds to a component of the electric or magnetic field of the light inside the fiber, $n(r)$ is the refractive index and $k_0 = \frac{2\pi}{\lambda_s}$ the wave vector of the laser radiation in vacuum. In order to solve this equation, we assume a wave propagating along the fiber axis with a propagation constant β as

$$\Psi(r, \phi, z) = \psi(r)e^{-il\phi}e^{-i\beta z}, \quad l = 0, \pm 1, \pm 2, \ldots, \tag{5.98}$$

where, we already took into account that the wavefunction needs to be unambiguous in the azimuthal direction, resulting in an azimuthal mode number l. Using the abbreviations $k_1^2 = n_{core}^2 k_0^2 - \beta^2$ and $k_2^2 = \beta^2 - n_{cladding}^2 k_0^2$, Eq. (5.97) can be transformed into a standard differential equation

$$\frac{\partial^2 \psi}{\partial r^2} + \frac{1}{r}\frac{\partial \psi}{\partial r} + \left(k_1^2 - \frac{l^2}{r^2}\right)\psi = 0, \quad r < a, \tag{5.99}$$

$$\frac{\partial^2 \psi}{\partial r^2} + \frac{1}{r}\frac{\partial \psi}{\partial r} + \left(k_2^2 + \frac{l^2}{r^2}\right)\psi = 0, \quad r > a. \tag{5.100}$$

These are the Bessel differential equations, which are solved by the Bessel functions J_l and K_l

$$\psi(r) \propto J_l(k_1 r), \quad r < a \tag{5.101}$$

$$\psi(r) \propto K_l(k_2 r), \quad r > a. \tag{5.102}$$

In the case of small refractive index differences $\Delta n = \frac{NA_{core}^2}{2n_{core}^2} < 0.01$, the modes guided by the fiber are linearly polarized [30]. They are therefore called **LP modes**. To describe these modes, the **fiber parameter**, **V parameter**, or **normalized frequency**

$$V = \frac{2\pi a NA_{core}}{\lambda}, \tag{5.103}$$

is used. Therein, $NA_{core} = \sqrt{n_{core}^2 - n_{cladding}^2}$ is the numerical aperture of the core. The fiber parameter can also be used to express the number of propagating modes M of the fiber, given by

$$M = \frac{4V^2}{\pi^2} + 2 \quad \text{for } V \gg 1. \tag{5.104}$$

Fig. 5.36 Intensity distribution of a Gaussian and a Bessel function

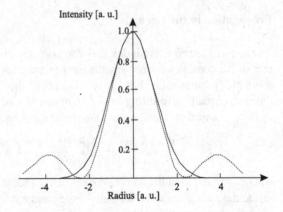

It can be shown that for $V < 2.405$, i.e. the zero of the lowest Bessel function, only the fundamental mode LP_{01} is guided in the fiber. These fibers are called monomode or single-mode fibers. However, it has to be taken into account that two LP_{01} modes exist which have perpendicular polarizations.

The transverse intensity profile of this fundamental mode LP_{01} is, however, very similar to the one of a Gaussian beam with power P as can be seen in Fig. 5.36, especially for the central lobe which is the one propagating in the fiber. Thus, we can assume a Gaussian distribution in the fiber core given by [48]

$$I(r) = \frac{2P}{\pi w_0^2} e^{-\frac{2r^2}{w_0^2}}, \tag{5.105}$$

for which the mode field radius can be calculated by the empiric formula

$$w_0 = a\left(0.65 + 1.619 V^{-1.5} + 2.876 V^{-6}\right). \tag{5.106}$$

Spectroscopic Properties of Fiber Lasers

In the following we will discuss the spectroscopic properties of fiber lasers. They differ from the properties of bulk solid-state lasers as a result of the influence of the waveguide effect in the fiber. As an example, we will investigate a Tm^{3+}:ZBLAN fiber laser. Comparing the overlap between absorption and emission in Fig. 5.37, it is evident that a Tm^{3+}-doped ZBLAN fiber laser can be operated as a laser or amplifier in a range between 1.85 μm to over 1.95 μm. Below this range the reabsorption will be too high and one would have to use small fiber lengths and thus small pump volumes. Above this range the gain becomes too low and the intrinsic losses thus result in high laser thresholds and low efficiencies.

In this context, it has to be noted that the emission cross section can only be determined either by measuring the fiber preform or by recording the fluorescence emitted perpendicularly from the outer fiber surface. This arises from the fact that the waveguiding along the fiber axis strongly changes the fluorescence spectrum of

Fig. 5.37 Cross sections of stimulated emission and absorption of Tm^{3+}:ZBLAN [31, 32]

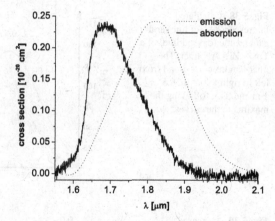

the guided light. To model this, we start from the radiation transport equation of the spectral signal power in a pumped medium

$$\frac{\partial \tilde{P}_s(z)}{\partial z} = \Gamma\big[\sigma_e N_2(z) - \sigma_a N_1(z)\big]\tilde{P}_s(z) + \sigma_e \Gamma \tilde{P}_0 N_2(z), \qquad (5.107)$$

wherein N_1 and N_2 are the population densities of the lower and upper laser level, respectively. The mode overlap factor between the guided Gaussian mode and the doped core is given by

$$\Gamma = 1 - e^{-\frac{2a^2}{w_0^2}}. \qquad (5.108)$$

Assuming that all ions are either in the ground state or the excited state so that $N_1(z) + N_2(z) = N_g = N_{Tm}$ is the total Tm^{3+}-doping density, we obtain for the amplification of a signal at a wavelength λ_s in a fiber of length L

$$G(\lambda_s) = e^{\int_0^L \Gamma[\sigma_e(\lambda_s)N_2(z') - \sigma_a(\lambda_s)N_1(z')]dz'}. \qquad (5.109)$$

We now assume that the maximum amplification of the fiber is G_{max} and occurs at a wavelength λ_{max}, which can be calculated from

$$\frac{\partial G(\lambda_s)}{\partial \lambda_s} = 0. \qquad (5.110)$$

The maximum gain wavelength thus solves the equation

$$\frac{\partial \sigma_e(\lambda_s)}{\partial \lambda_s} = \frac{\partial \sigma_a(\lambda_s)}{\partial \lambda_s}\left(\frac{\sigma_e(\lambda_s) + \sigma_a(\lambda_s)}{\frac{\ln G_{max}}{\Gamma N_g L} + \sigma_a(\lambda_s)} - 1\right). \qquad (5.111)$$

In this equation, only the spectroscopic properties of the fiber and its length occur. It is therefore independent of the real axial repartition of the ion densities $N_i(z)$. The cross sections are then expressed by a sum of Gaussian functions and a numeric root finding algorithm can be used to obtain λ_{max}.

However, in order to obtain the fluorescence spectrum at the fiber end, we have to integrate the radiation transport equation including spontaneous emission. For

Fig. 5.38 Calculated fluorescence solid line) and gain profile (dashed line) of a Tm^{3+}:ZBLAN fiber. The fiber lengths correspond from left to right to 0.3 m, 0.5 m, 1 m and 3 m, following the maxima in the curves

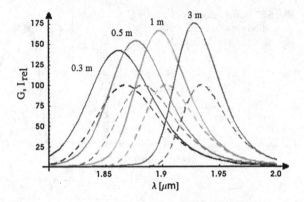

Fig. 5.39 Measured fluorescence of a Tm^{3+}:ZBLAN fiber under excitation at 792–805 nm for different fiber lengths L

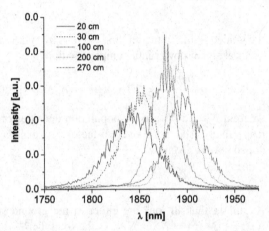

simplicity, we assume a homogeneous distribution of the excitation density. This results in a relative fluorescence intensity of

$$I_{rel}(\lambda) = \frac{\sigma_e(\lambda)}{\sigma_e(\lambda) + \sigma_a(\lambda)} \big(G(\lambda) - 1\big)\left(1 + \frac{\sigma_a(\lambda) N_g \Gamma L}{\ln G(\lambda)}\right), \qquad (5.112)$$

with

$$G(\lambda) = e^{\frac{\sigma_e(\lambda) + \sigma_a(\lambda)}{\sigma_e(\lambda_{max}) + \sigma_a(\lambda_{max})}[\ln G_{max} + \Gamma N_g L \sigma_a(\lambda_{max})] - \Gamma N_g L \sigma_a(\lambda)}. \qquad (5.113)$$

This simple relation only holds for axially constant population densities.

The result of the simple calculation is shown for a fiber with $\Gamma = 0.788$ and $N_g = 3.95 \times 10^{26}$ m^{-3} at a maximum amplification of $G_{max} = 100$ in Fig. 5.38. It shows a shift of the maximum of the fluorescence with increasing fiber length and a simultaneous decrease in fluorescence bandwidth. These effects arise from the re-absorption, which increases with increasing fiber length.

The experimental results can be seen in Fig. 5.39 and correspond well to the theory regarding the wavelength shift. However, in this experiment the maximum gain was probably not identical for all fiber lengths, causing the differences in absolute values.

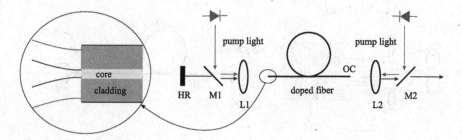

Fig. 5.40 Schematic of the experimental setup of a fiber laser

Experimental Arrangements of Fiber Lasers

The principle experimental arrangement of a double-clad fiber laser can be seen in Fig. 5.40. For a core-pumped fiber laser, this setup only differs in the fact that the pump light needs to have a better beam quality and that the cladding has a lower index of refraction compared with the core. In order to obtain an evenly distributed excitation and heat load, the fiber is usually pumped from both ends. In many laboratory configurations, often free-space resonators are still used to have a maximum number of adjustable parameters in the arrangement and the possibility to include various elements into the cavity. Here, dichroic mirrors (M1 and M2) are used to combine the pump radiation with the laser mode. The pump beams are then launched into the fiber by two lenses (L1 and L2), which simultaneously act as collimation lenses for the laser mode. On one side, the laser beam is retro-reflected by an external highly-reflecting mirror (HR), forming the **external cavity**. Owing to the high gain achieved in the long fiber, a large amount of outcoupling can be used. Therefore, it is often sufficient to simply use the perpendicularly-cleaved fiber end on the other side of the fiber as the OC mirror. This provides a Fresnel reflectivity of

$$R_{OC} = \left(\frac{n-1}{n+1}\right)^2, \tag{5.114}$$

which is around 3.4 % for silica fibers with a refractive index of $n \approx 1.45$. One advantage of this external-cavity design is the free-space propagation of the intracavity beam, which enables the insertion of modulators for Q-switching or of frequency-selective elements such as etalons for wavelength selection and tuning. However, the external-cavity configuration has one drawback, which is the increase in intracavity losses due to the coupling losses that occur when the laser light is re-injected into the fiber core. As this re-injection efficiency strongly depends on the mode matching between the reflected beam from the HR mirror and the guided mode of the fiber, the external cavity has to be well designed and aligned.

In order to avoid this adjustment problem in cw high-power lasers, an all-fiber solution can be produced, as shown in Fig. 5.41. Pump combiners are used to couple the pump light into the cladding. These combiners consist of several smaller

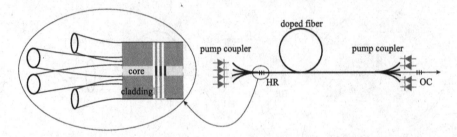

Fig. 5.41 Schematic of the experimental setup of a fiber laser in an all-fiber design without external components

Fig. 5.42 Schematic of the experimental setup of a grating-tuned fiber laser

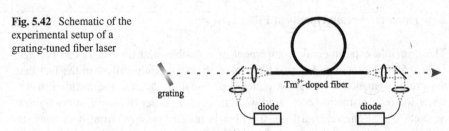

undoped multi-mode fibers which are **spliced**, i.e. welded, to the cladding of an undoped double-clad fiber that matches the doped laser fiber in diameter and index profile. Thus, the core of this passive fiber is a freely accessible port for the configuration of the cavity, whilst the smaller pump fibers are connected to high-power laser diodes for pumping. Finally, this undoped **combiner** is spliced to the doped laser fiber.

To obtain a fully self-contained, i.e. all-fiber design, fiber laser without any intracavity free-space propagation, **Bragg gratings** are written into both ends of the fiber core to act as HR and OC mirror. These Bragg gratings are a periodic refractive index structure similar to dielectric mirrors. However, they are not created by a thin-film deposit. This refractive index pattern is written into the fiber by illumination of the fiber with a UV-laser-created interference pattern, e.g. by using an Ar^+-ion laser, or by a femtosecond laser. Such Bragg gratings have very sharp resonances with high reflectivity and are written for a specific wavelength. However, slight tuning is possible, e.g. by heating or cooling of the Bragg grating, or by stretching the grating-containing part of the fiber.

Examples of a CW Fiber Laser Based on Tm^{3+}

As an example, Fig. 5.42 shows the experimental arrangement of a Tm^{3+}:ZBLAN fiber laser, which is pumped from both sides by two fiber-coupled laser diodes at 792 nm [2]. The fiber has a core diameter of 30 µm, NA $= 0.08$, and a cladding diameter of 300 µm, NA $= 0.47$. In this setup the cavity was built by a retro-reflective mirror (or grating) and the other polished fiber end, i.e. with an OC reflectivity of

Fig. 5.43 Measured output power of a Tm^{3+}:ZBLAN fiber laser tuned to different wavelengths

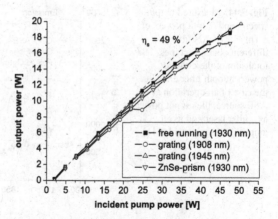

$R \sim 4\,\%$. The emission wavelength was coarsely set by the fiber length and the optimum pump absorption was adjusted by arranging for an appropriate ratio between core and cladding area. The experimental performance can be seen in Fig. 5.43. As expected by the long upper-state lifetime and the negligible multi-phonon relaxation in ZBLAN visible in Table 5.2, the threshold is very low and due to the efficient cross-relaxation mechanism a high slope efficiency of 49 % with respect to the launched pump power is achieved. At high pump powers a **thermal roll-over** occurs, which is not connected to thermal lensing but results from the heating of the fiber, which was uncooled in this arrangement [35].

An interesting effect is the wavelength independence of the laser efficiency over a broad wavelength range, which was also confirmed for different modes of operation of the fiber, used as a fiber laser, an amplifier or as a free-running amplified-spontaneous-emission (ASE) source. The arrangements operated at completely different wavelengths and therefore at different emission cross sections and reabsorption levels as can be seen in Fig. 5.44.

This effect occurs as a result of the amorphous nature of the glass host and shows that an inhomogeneously broadened laser medium can react as quasi-homogeneously broadened. In the glass host the crystal field varies between different positions in the fiber, resulting in a position-dependent Stark-level splitting and energy shift for the Tm^{3+} ions in the fiber. However, these site-to-site shifts in the energy levels correspond approximately to the Stark-splitting within the manifolds. Therefore, a given wavelength within the large emission band can interact with nearly all the Tm^{3+} ions in the fiber, connecting however different levels in each ion. In this way, a given wavelength can extract energy from nearly all of the ions and the medium acts quasi-homogeneously. Only at the lower and upper limit of the emission range is this effect reduced resulting from a lower possibility of transitions that match the wavelength.

This quasi-homogeneous-broadening effect is also found in Tm^{3+}:silica, as can be seen in Fig. 5.45. A Tm^{3+}-doped silica fiber with a core diameter of 20 µm, NA = 0.2, and a cladding diameter of 300 µm, NA = 0.4 has been used. The optimum fiber length was found to $L = 2.3$ m from the measurements also shown in Fig. 5.45.

Fig. 5.44 Measured output
spectra and output powers of
a Tm^{3+}:ZBLAN fiber in
different operation types. The
total sum of the emitted
powers at both fiber ends in
the case of an operation as an
ASE source, the output power
as a fiber laser and as an
amplifier is comparable [41]

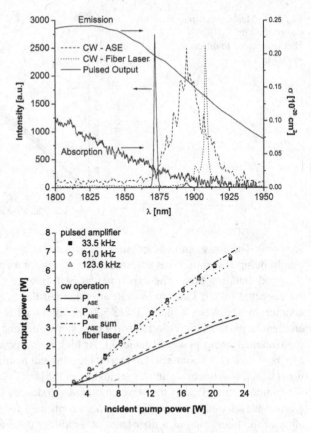

Q-Switched Fiber Lasers

In contrast to low repetition rate Q-switched operation of, e.g. a Tm^{3+}:YAG laser
[33], in which the maximum laser efficiency is determined by the ratio between the
pump pulse duration and excited state lifetime given by Eq. (4.6), high-repetition
rate ($\nu_{Rep} \gg \frac{1}{\tau}$) Q-switched Tm^{3+} fiber lasers can be as efficient as in cw operation
[2, 34, 36, 37]. In Tm^{3+}:ZBLAN, owing to its long upper level lifetime, a low intra-
cavity laser intensity is sufficient to saturate the laser transition. Thus single-pass
amplifiers can be easily realized, reaching the same efficiency as in continuous-wave
laser operation [38–42]. However, as a result of the very high laser signal saturation
intensity in Tm^{3+}:silica, cavity losses, as for example the re-injection efficiency in
an external cavity, need to be minimized to get highly efficient operation [2].

Especially for pumping applications, in which Q-switched fiber lasers are
e.g. used to pump non-linear converters, and where high average output powers are
needed, the high repetition rate fiber lasers lead to stable and compact system archi-
tectures. As a result of the guiding effect in fibers, a strong influence of amplified
spontaneous emission (ASE) on the laser properties exists, and it can be shown that
ASE becomes a non-negligible effect when the amplification along the fiber exceeds

Fig. 5.45 Measured output power of a Tm^{3+}:silica fiber laser versus wavelength and fiber length

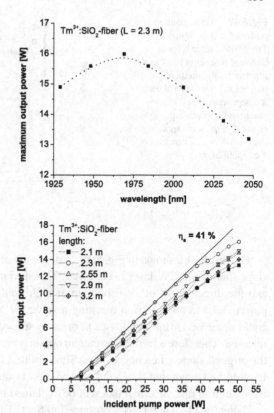

a factor of 100, i.e. 20 dB. However, as recent investigations on a Tm^{3+}:silica fiber laser have shown, ASE effects can be efficiently avoided in Q-switched fiber lasers at high repetition rates through good design [36]. The experimental arrangement is shown schematically in Fig. 5.46. An AOM is inserted into the external cavity for Q-switching and a telescope increases the beam diameter on the end mirror, where there is a diffraction grating for wavelength tuning, to avoid optical damage at high pulse energies. Also EOM cavity blocking can be used; however, as this fiber is not polarization maintaining, an AOM provides much lower insertion losses than an EOM and a polarizer.

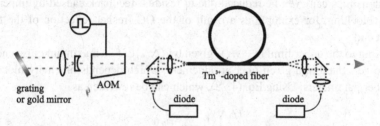

Fig. 5.46 Schematic of the experimental setup of an acousto-optically Q-switched fiber laser

Fig. 5.47 Average output power of a Q-switched Tm^{3+}:silica fiber laser at different repetition rates as a function of the incident pump power [36]. The inset shown the dependence of the maximum allowable pump power and the corresponding output power as a function of the repetition rate

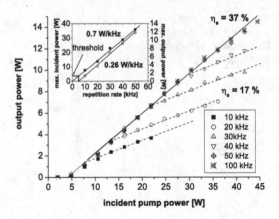

The results are shown in Fig. 5.47. A deviation of the output power from the 37 % slope line of the CW laser is observed. This deviation is repetition rate dependent and the direct result of ASE buildup at high pump intensities. These high pump powers lead to an inversion reaching a value for which the intracavity modulator could not generate enough losses in order to suppress ASE before the next pulse is released. Therefore a further increase in output power is only connected to the half of the original slope efficiency as the ASE is emitted from the fiber in both directions. It could be shown that this point of deviation in output power is linearly linked to the repetition rate, an effect that will be explained in the following discussion.

Using the Q-switch theory presented in Sect. 4.1.1 for a repetitively Q-switched laser, the initial inversion increases with pump power and is limited by the maximum population inversion $\langle \Delta N \rangle_\infty$, which would be reached at the incident pump intensity $I_{p,0}$ in the blocked cavity without a laser signal present for $t \to \infty$, given by

$$\langle \Delta N \rangle_\infty = \frac{2\lambda_p \tau}{hc} \frac{\eta_{abs}}{L} I_{p,0} - \langle N \rangle. \qquad (5.115)$$

However, even in the blocked cavity $\langle \Delta N \rangle_\infty$ will never be reached as the buildup of ASE will limit the inversion to a lower value $\langle \Delta N \rangle_\infty^{ASE}$, which corresponds to the ASE threshold $I_{p,0}^{ASE}$ of the fiber. This pump intensity at which ASE dominates the fiber dynamics can even be reduced due to residual feedback caused by imperfect cavity blocking, for example as a result of the OC fresnel reflection of the fiber output end.

Owing to the upper limit in $\langle \Delta N \rangle_i$ given by $\langle \Delta N \rangle_\infty^{ASE}$, a maximum pulse energy can be generated in a given fiber arrangement, which depends on cavity blocking and fiber parameters. Using Eq. (4.28), which can be rewritten as

$$\frac{\langle \Delta N \rangle_f'}{\langle \Delta N \rangle_i'} = e^{-r\eta_e(r)}, \qquad (5.116)$$

in which r is the ratio between pump and threshold power and $\eta_e(r)$ the extraction efficiency, and Eq. (4.57), the initial population inversion in repetitive Q-switched operation can be written as

$$\langle \Delta N \rangle_i' = \langle \Delta N \rangle_\infty' \frac{1 - e^{-\frac{1}{\tau v_{Rep}}}}{1 - e^{-r\eta(r)}e^{-\frac{1}{\tau v_{Rep}}}}, \tag{5.117}$$

which simplifies to

$$\langle \Delta N \rangle_i' = \langle \Delta N \rangle_\infty' \left(1 - e^{-\frac{1}{\tau v_{Rep}}}\right) \tag{5.118}$$

for $r \gg 1$, see Fig. 4.6. At high repetition rates $v_{Rep} \gg \frac{1}{\tau}$, this expression can be approximated and yields

$$\langle \Delta N \rangle_i' \approx \frac{\langle \Delta N \rangle_\infty'}{\tau v_{Rep}} \propto \frac{P_{p,0}}{\tau v_{Rep}} \tag{5.119}$$

in the case of low ground-state depletion. Owing to the upper limit in $\langle \Delta N \rangle_i$ given by $\langle \Delta N \rangle_\infty^{ASE}$, the point of deviation from linearity corresponds to a linear relation between the maximum allowable pump power $P_{p,0}^{max}$ and the repetition rate,

$$P_{p,0}^{max} = k_p v_{Rep}. \tag{5.120}$$

The linear slope relation between pump and output power then causes a corresponding relation between the laser output power and the repetition rate, given by a factor k_s that can be derived as

$$k_s = \frac{\partial P_{out}(P_{p,0}^{max})}{\partial v_{Rep}} = k_p \eta_s, \tag{5.121}$$

where, η_s is the laser slope efficiency. These theoretical predictions well agree with the experiment as can be seen in Fig. 5.47. The corresponding proportionality factors found were $k_p = 0.7 \frac{W}{kHz}$ and $k_s = 0.26 \frac{W}{kHz}$ for incident pump powers up to 45 W. This shows that under high-repetition-rate operation, high average powers can be achieved with Q-switched fiber lasers.

To maximize the generated pulse energy a careful laser design is needed to increase the ASE threshold of the blocked cavity, allowing high pump powers above laser threshold, i.e. high values of r, which then result in very short pulses according to Eq. (4.30) and Fig. 4.6. This was also verified experimentally, where short pulses of 41 ns could be achieved from a 2.3 m long Tm^{3+}-doped silica fiber at $r = 15$ [36]. As a comparison, Q-switch theory results in a pulse width of 30 ns for this fiber, taking into account a cavity round-trip time of $\tau_{RT} \approx 22$ ns, which can be seen as the cavity photon lifetime resulting from the strong outcoupling. This shows that in order to obtain short pulses and to push the ASE threshold as high as possible, a very low OC reflectivity has to be used in Q-switched fiber lasers.

Power Limits in Fiber Lasers

The fact that the laser mode is concentrated in the small core in a fiber laser causes three different constraints in output power, depending on the spectral and temporal properties of the laser radiation. The first limit is connected to the finite optical-damage threshold of silica glass, which is on the order of 3 GW/cm^2. This is the intensity, at which a damage is created on the surface of the glass, usually together with the formation of a plasma and the vaporization of parts of the glass surface. For the example of a 30 μm-diameter core, NA $= 0.04$ fiber at a wavelength around 1 μm, a mode field radius of $w_0 = 13$ μm is calculated from Eq. (5.106) and results in a damage threshold of

$$P_{max} = \frac{\pi w_0^2}{2} \hat{I} = 8 \text{ kW}. \tag{5.122}$$

At this power, damage to the fiber end faces can occur, making it necessary to re-cleave or to repolish the fiber ends. It is important to note that this limit gives an instantaneous power, i.e. it is independent of the temporal mode of operation of the fiber laser and has the same value for a cw fiber laser, where it denotes the laser power, and for a pulsed Q-switched fiber laser, where it denotes the maximum pulse peak power at which damage is likely to occur.

The two other processes limiting the output power of a fiber laser are Brillouin and Raman scattering, two intensity-dependent non-linear processes.

Brillouin scattering   In Brillouin scattering, a photon of the laser field propagat-ing in the fiber is scattered by an acoustic phonon. Consequently, the most efficient scattering occurs with longitudinal acoustic phonons. Is a phonon created during that process, the laser photon gets red-shifted by an amount of ν_B, which is a function of the fiber material and the scattering angle between the photon and the phonon. This is called the **Stokes process**. In the reverse case, i.e. when a phonon is anni-hilated in the scattering process, the photon gets blue-shifted, which is called the anti-Stokes process. The maximum frequency shift is obtained in reverse scattering, in which the scattered photon propagates in the opposite direction than the incoming unscattered photon. This maximum frequency shift is given by [43]

$$\nu_B = \frac{2n v_s}{\lambda_0}, \tag{5.123}$$

where, v_s denotes the velocity of sound of the longitudinal phonons, n the refrac-tive index of the host medium and λ_0 the wavelength of the incoming, unscat-tered radiation, the so-called **Brillouin pump radiation**. For ZBLAN this results in $\nu_B \approx 18.82$ GHz [45].

The interference between the back-scattered light and the incident pump radia-tion creates an intensity pattern. When the spatial period of this pattern corresponds to the phonon wavelength and propagates itself with the velocity of sound in the medium, the lattice deformation created by the phonon will be amplified by **elec-trostriction**. This feedback increases the Brillouin-scattering rate and is called **stim-ulated Brillouin scattering** (SBS). Thus, a threshold exists and for powers above a strong conversion from the incoming radiation into the scattered radiation occurs. However, Brillouin scattering can only occur if the optical pulse width is longer than the average photon lifetime of the medium, which is $\tau_{ph} = 3.3$ ns for the case of ZBLAN.

Table 5.4 SBS thresholds in ZBLAN at 1.87 μm. The real thresholds are calculated for a 0.23 nm-wide signal line.

	Fiber 1	Fiber 2	Fiber 3
L	0.3 m	1.5 m	2.4 m
w_0	3.70 μm	7.2 μm	12.8 μm
$P_{SBS,0}$	602 W	456 W	900 W
P_{SBS}	107 kW	81.2 kW	160 kW

In order to derive an expression for the Brillouin threshold, the Brillouin amplification of a small-line-width pump radiation $\Delta v \ll \Delta v_B$ is used, given by

$$g_B = \frac{2\pi n^7 p_{12}^2}{\Delta v_B c \lambda^2 \rho v_s}, \tag{5.124}$$

where, $\Delta v_B = \frac{1}{\pi \tau_{ph}}$ is the Brillouin line width caused by the natural phonon lifetime (natural line width), ρ the density of the fiber medium and p_{12} its elasto-optic coefficient. The frequency-dependent small-signal gain of the SBS shows a Lorentzian line shape and can be described by

$$g_{SBS}(v) = g_B \frac{\frac{\Delta v_B^2}{4}}{(v - \frac{c}{\lambda_0} + v_B)^2 + \frac{\Delta v_B^2}{4}}, \tag{5.125}$$

and the SBS threshold power for a low line width pump radiation results in [43]

$$P_{SBS,0} \simeq 21 \frac{A_{eff}}{g_B L_{eff}}, \tag{5.126}$$

where, the effective fiber length

$$L_{eff} = \frac{1}{\alpha}\left(1 - e^{-\alpha L}\right) \tag{5.127}$$

is nearly equal to the geometrical fiber length L owing to the low intrinsic losses α in the glass medium. The effective mode area is given by $A_{eff} = \frac{\pi w_0^2}{2}$ and can be approximated in multi-mode fibers by the core area $A = \pi a^2$.

However, if the line width of the laser radiation is much larger than the Brillouin line width of the glass, e.g. $\Delta v_B = 96$ MHz in ZBLAN, the Brillouin line width Δv_B in Eq. (5.124) has to be replaced by the laser line width Δv [44], resulting in a real Brillouin threshold of

$$P_{SBS} \simeq 21 \frac{A_{eff}}{g_B L_{eff}} \frac{\Delta v}{\Delta v_B}. \tag{5.128}$$

An example for SBS thresholds is given in Table 5.4. It can be seen, that in contrast to the low thresholds obtained for a small laser line width, a fiber laser or amplifier operating at a line width on the order of 0.1–1 nm shows Brillouin thresholds that are much higher than the optical damage threshold of a standard fiber for short fiber lengths. However, for long fibers the Brillouin threshold can be much lower than the optical damage threshold and has to taken into account in high power laser systems.

Raman scattering   In analogy to Brillouin scattering a photon can also scatter on optical phonons, which is called **Raman scattering**. The Raman effect, however, is different from the Brillouin effect in several points. First, the frequency shifts are much larger owing to the higher phonon energies of optical phonons, causing a frequency shift ν_R in the range of some THz, where the red-shifted radiation is also called the **Stokes radiation**. And second, the decay times of the optical phonons are much shorter than those of the acoustic phonons, causing Raman scatting also to occur at laser pulse widths smaller than 1 ns.

In **stimulated Raman scattering (SRS)** a scattering in the same propagation direction as well as in the reverse direction is possible. However, here the lowest threshold is obtained for scattering into the same direction of propagation, resulting in

$$P_{SRS} \simeq 16 \frac{A_{eff}}{g_R L_{eff}}. \tag{5.129}$$

The threshold for reverse scattering is about 25 % higher [43], and thus does not need to be taken into account here. By passing this threshold, a large amount of the incident radiation becomes red-shifted with a high efficiency. In principle, this process can repeat itself for the created Stokes radiation, causing successive Stokes orders exciting the fiber.

The Raman amplification is given by

$$g_R = \frac{4\pi \chi_R''}{\lambda_{st} n^2 \epsilon_0 c}, \tag{5.130}$$

wherein χ_R'' denotes the non-linear susceptibility of the glass medium, and λ_{st} the wavelength of the Stokes-shifted light. In glasses, depending on the chemical composition of the glass, several maxima can occur in the density of states of the phonons, e.g. for ZBLAN one obtains 17.7 THz (590 cm^{-1}), 14.4 THz (480 cm^{-1}), 11.7 THz (390 cm^{-1}), 9.9 THz (330 cm^{-1}), 8.1 THz (270 cm^{-1}) and 6.0 THz (200 cm^{-1}) [46]. The strongest Raman amplification occurs on the 590 cm^{-1} line. In ZBLAN, for example, the Raman gain is about 21 THz wide. Thus, the Raman pump signal, i.e. the laser radiation, can be seen as quasi-monochromatic compared to this large Raman line width, and the threshold does not need to be scaled as we did for the Brillouin scattering.

As an example, Table 5.5 shows the corresponding SRS thresholds for the three fibers of Table 5.4.They are much closer to the optical-damage limit and usually determine the upper laser power limit for long fibers, especially in pulsed operation.

The only possibility to avoid these non-linear effects is the reduction of the laser intensity in the fiber by increasing the mode-field diameter. However, in order not to loose the modal properties of the fiber, the fiber parameter in Eq. (5.103) has to stay constant. As the numerical aperture of a step-index fiber usually has a lower limit of NA ≥ 0.04 as a result of the fiber-manufacturing process, the core diameter

Table 5.5 SRS thresholds for three ZBLAN fibers at 1.87 μm.

	Fiber 1	Fiber 2	Fiber 3
L	0.3 m	1.5 m	2.4 m
w_0	3.70 μm	7.2 μm	12.8 μm
P_{SRS}	26.4 kW	20.0 kW	39.5 kW

is limited to ∼30 μm. Even if lower numerical apertures would be possible, high bending losses would result and the fiber laser would loose its beneficial property of being coiled to obtain a small laser volume. These upper limits in core size are related to the principle of the step-index fiber. Using **photonic-crystal fibers** the effective core diameter can be greatly increased. The guiding in these fibers is not caused by a step in the refractive index, but by a wave-optical effect: The core is surrounded by an air-filled hole pattern, which results in a band-structure for the light frequencies, comparable with the energy band-structure of electrons in a crystalline solid. Thus, for certain wavelength bands, a band-gap exists, and these wavelengths are not allowed to propagate in the structure around the core. This confines the light to the core area. Another simple argument to explain this guiding effect is that by introducing the air holes, the average refractive index of the cladding is lower than the core. However, this simple argument does not allow to explain the frequency-spectrum of the guiding band gaps. These photonic-crystal fibers allow single-mode operation with core diameters of over 100 μm. However, a large core diameter with single-mode guiding corresponds to a small NA fiber. Thus, very high bending losses occur and these fibers have to be aligned straight in order to avoid these losses.

Using these photonic-crystal fibers, single-mode CW output powers of several kW have been realized with Yb^{3+}-doped silica fibers.

Applications

Most high power fiber lasers are Yb^{3+}-doped silica fibers emitting in the 1.03–1.08 μm range and are used in welding and cutting applications. In 2006, the state-of-the-art was 2 kW in a single-transverse mode realized by *IPG Photonics, Burbach, Germany*, while multi-mode fiber lasers generate >10 kW out of a 100–300 μm-core multimode fiber. These systems reach efficiencies of up to 25 %. However, the fiber laser units usually consist of several modules emitting some 100 W, which are then all coupled into a single 100–300 μm-core undoped transport fiber. Due to the low beam quality of the multimode output fiber, these sources can only be used with short distances between the fiber output optics and the workpiece. In 2012, single-transverse-mode fiber lasers of 10 kW output power have been produced—a value close to the damage threshold calculated above. Other important applications are in medicine and arise from the easy delivery of the laser radiation with the fiber, that can be introduced into endoscopes for low-invasive surgery. In these applications, however, a radiation around 2 μm is better suited due to its strong absorption in water.

Fig. 5.48 Schematic of a thin-disk-laser setup, showing the laser disk, which is soldered to the heat sink on its HR side using indium solder, and the external OC mirror [47]

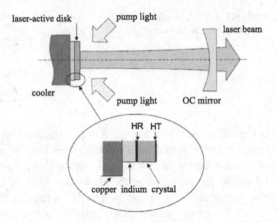

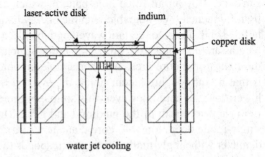

Fig. 5.49 Mechanical design of the disk heat sink with water-jet cooling [47]

5.3.3 The Thin-Disk Laser

The basic idea behind the thin-disk laser is an axial, one-dimensional heat flow within the laser medium towards the heat sink. Therefore, at any axial point along the laser medium, a homogeneous radial temperature distribution is predicted to occur and consequently, no thermal lensing results. The schematic of such a laser is depicted in Fig. 5.48. The laser-active medium is a disk with a diameter of several mm to some cm, with a thickness of some 100 µm. The disk is AR coated for a high transmission at its front face, and HR coated at its back face, with which it is soldered to the copper heat sink using indium or gold solder. The laser cavity is formed by the HR coating on the disk and an external OC mirror. This mirror allows control of the cavity mode size to match the pumped area in the disk.

The heat-sink assembly of the disk is shown in Fig. 5.49. The laser-active disk is soldered with indium on to a larger copper disk, which itself is mounted on to a hollow copper block. Inside this block, a water jet is emitted that hits the back surface of the copper disk for efficient cooling. Water-jet cooling is chosen as it is much more efficient than a simple laminar flow along the copper surface.

The pumping scheme of a disk laser differs from usual longitudinal pumping, as the disk itself shows a small single-pass absorption as a result of the low thickness of the disk. In order to enhance the pump absorption, the pump beam makes multiple passes through the disk. This multi-pass pumping scheme is shown in Fig. 5.50.

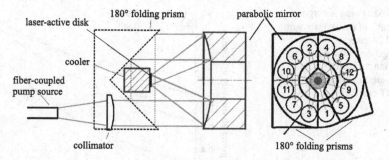

Fig. 5.50 Schematic of the pump system of a disk laser, using a parabolic mirror and retro-reflecting prisms to achieve a total 24-pass ($m = 12$) propagation of the pump radiation [47]

The pump light, usually emitted by a highly-multi-mode fiber or a glass-rod homogenizer, is collimated by a first lens and directed on to a parabolic mirror, which images the fiber output on to the disk. The non-absorbed pump power then is reflected from the disk, hits the parabolic mirror on the opposite radial point and is collimated again. Around the disk-heat-sink assembly, two 180° retro-reflecting prisms are used to flip the residual pump beam propagation path and to direct it on to another spot on the parabolic mirror to perform a second pass through the disk. After being re-collimated by the parabolic mirror, the residual pump light after the second pass hits the other 180° retro-reflecting prism, which is tilted with respect to the first prism. This optical configuration causes the pump beam to make multiple passes through the disk. The optical path lengths in this pump arrangement are chosen so that the disk is imaged on to itself after each pump pass (**relay imaging**).

The theoretical number of passes only depends on the tilt angle between the two prisms. After the first m passes, where m denotes the number of passes through one disk thickness L, the beam hits one of the prisms symmetrically on the reflection axis, causing the beam to be retro-reflected. Therefore, this multi-pass pump optic achieves in total $2m$ pump passes through the disk. The corresponding tilt angle between the two prisms is thus

$$\gamma = \frac{360°}{2m}, \tag{5.131}$$

i.e. 15° for the $2m = 24$ pass setup shown in Fig. 5.50.

The effective number of pump passes through the disk, however, depends strongly on the pump absorption of the disk and the values of the reflectivity of the mirror, prism and disk coatings and the transmission of the AR coating on the disk. Assuming a single-pass disk transmission of

$$T_D = e^{-\alpha_p L}, \tag{5.132}$$

with α_p being the pump absorption coefficient and L the disk thickness, we obtain a residual power after an even number of m passes, i.e. after a number of $\frac{m}{2}$ back-and forth passes (reflections) through the disk, of

$$P_{res,m} = \left(R_P^2 R_{PR}\right)^{\frac{m}{2}-1} \left(T_{HT}^2 R_{HR} T_D\right)^m R_P P_{inc}, \tag{5.133}$$

Fig. 5.51 Pump absorption efficiency of a disk laser with a total 24-pass ($m = 12$) propagation of the pump radiation for various single-pass disk transmissions and HT coatings

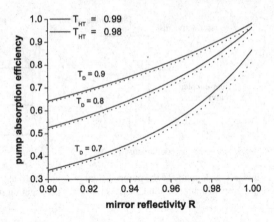

where, P_{inc} is the incident pump power entering the disk-laser module, R_P the reflectivity of the parabolic mirror, T_{HT} the transmission of the anti-reflection coating on the disk, R_{HR} the reflectivity of the HR coating on the disk for the pump light and R_{PR} the total reflectivity of a prism for one $180°$ turn. Thus, after the full $2m$ passes the residual pump power is given by

$$P_{res,2m} = \left(R_P^2 R_{PR}\right)^{m-1} \left(T_{HT}^2 R_{HR} T_D\right)^{2m} R_P\, P_{inc}. \tag{5.134}$$

In analogy, the amount of pump power absorbed during the $\frac{m}{2}$th reflection on the disk is given for an even m by

$$P_{abs,m} = (1 - T_D)(1 + T_D R_{HR})\left(R_P^2 R_{PR}\right)^{\frac{m}{2}-1}\left(T_{HT}^2 R_{HR} T_D^2\right)^{\frac{m}{2}-1} R_P\, P_{inc}. \tag{5.135}$$

Summing up all the contributions from the different passes, we obtain the total absorbed pump power as

$$P_{abs,2m}^{tot} = \sum_{k=1}^{m} P_{abs,2k}, \tag{5.136}$$

resulting in

$$P_{abs,2m}^{tot} = R_P(1 - T_D)(1 + T_D R_{HR})\frac{1 - (R_{HR} R_P^2 R_{PR} T_D^2 T_{HT}^2)^m}{1 - R_{HR} R_P^2 R_{PR} T_D^2 T_{HT}^2}\, P_{inc}. \tag{5.137}$$

In order to see the effect of the different reflectivities and transmission values on the total pump absorption, we assume a constant reflectivity value for all the different reflective surfaces of $R_{HR} = R_P = R_{PR} = R$. This results in a pump absorption efficiency of

$$\eta_{abs,2m}^{tot} = R(1 - T_D)(1 + T_D R)\frac{1 - (R^4 T_D^2 T_{HT}^2)^m}{1 - R^4 T_D^2 T_{HT}^2}, \tag{5.138}$$

which is shown for different values of T_D and T_{HT} for the case $m = 12$ in Fig. 5.51. It can be seen that for an efficient pump absorption, very good, highly-reflective, mirror coatings are necessary, and a very good anti-reflection coating on the disk

Fig. 5.52 Schematic of a multi-disk setup with a single cavity, in which the total single-pass gain is increased

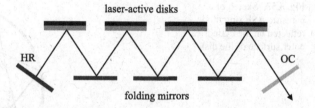

has to be used. These coatings are especially critical as they have to provide this performance at all the different incident angles occurring for the pump beams, taking into account that the pump light is usually unpolarized after the homogenization in the fiber or glass rod.

Another important point arises from the low gain resulting from the low thickness of the disk. Therefore, a high OC reflectivity has to be used and the intracavity losses have to be minimized to very low values in order to obtain a high laser efficiency. To increase the single-pass gain, several discs can be arranged within one laser cavity as shown in Fig. 5.52. This cavity consists of an OC and a HR mirror at the ends, and a zig-zag path between the different discs using passive folding mirrors. Using this multi-disk design, Yb^{3+}:YAG thin-disk lasers with output powers of >10 kW have been realized for welding and cutting applications in industry.

Power Limitation in Disk Lasers

A main limitation of the performance of disk lasers is the maximum extractable power per disk. This is a function of the disk size and is limited by amplified-spontaneous emission (ASE) mainly occurring in the transverse direction in the disk, where the gain-length product is much larger than in the axial resonator-mode direction. This aspect is especially important for closed paths within the disk, for which at each intersection with the disk boundary a total-internal reflection occurs, the round-trip losses on the path can become very low, resulting in a self oscillation of the disk on the internally trapped modes, called **parasitic modes**. At high OC transmissions, which are used in high-power lasers, these modes may therefore reach their threshold at a pump power much lower than the threshold of the cavity modes. Especially for power scaling, in which in principle an increase of the disk diameter at constant pump intensity should yield very high output powers, the onset of parasitic lasing can set an upper limit on the disk size and output performance of the laser. In a circular disk laser, three different parasitic modes can occur, depending on their main propagation schemes: ring modes, transverse modes and radial modes.

The ring modes are oscillations that are only reflected on the cylindrical outer surface of the disk. Assuming that an efficient reflection will only occur at incident angles θ_r on that surface, which are larger than the critical angle of total-internal reflection

$$\theta_c = \arcsin \frac{1}{n} \tag{5.139}$$

Fig. 5.53 Sketch of a
parasitic ASE ring mode
reflected on the cylindrical
outer surface of the disk

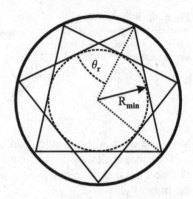

Fig. 5.54 Transverse mode
propagation through the disk
center

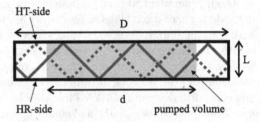

in the laser medium, we directly obtain from Fig. 5.53 that these modes cannot pass
an inner disk region with a radius R_{min} given by

$$R_{min} = \frac{D}{2}\sin\theta_c = \frac{D}{2n},\tag{5.140}$$

where, D is the diameter of the disk and n its refractive index. This relation, is
therefore, the key to the suppression of the ring modes in a disk laser. The radius
of the pumped area just needs to be smaller than the minimum ring-mode distance
from the disk center, given by R_{min}, and possible ring modes do not exhibit a gain
during their propagation. In the case of a quasi-three-level laser medium like Yb^{3+}-
doped crystals, the strong reabsorption on the laser line in the unpumped part creates
an additional loss, helping to suppress the ring modes further. Also, a hybrid disk
can be used, in which the outer part is doped with a strongly absorbing ion.

The transverse modes are reflected on the front and back surface of the disk and
on the cylindrical part. The maximum gain per mode round trip is obtained for the
transverse mode, which propagates through the disk center, shown in Fig. 5.54. Ac-
counting for the angle-dependent reflectivities of the three disk surfaces, this mode
should start oscillating when the gain along the propagation path compensates for
the reflection losses, i.e. when

$$R_{HR}^k(\theta_a)R_{HT}^k(\theta_a)R_C(\theta_a)e^{\frac{gD}{\sin\theta_a}} = 1,\tag{5.141}$$

where, $R_{HR}(\theta_a)$, $R_{HT}(\theta_a)$ and $R_C(\theta_a)$ are the angle-dependent reflectivities of the
HR, HT and cylindrical side of the disk, and k is the number of reflections, given by

$$k = \frac{D}{2L\tan\theta_a}.\tag{5.142}$$

Taking into account that the gain-diameter product gD in Eq. (5.141) accounts for the average gain of the mode propagating from one side to the other, and assuming that only an inner diameter d of the disk is pumped to suppress the ring modes, this average gain-diameter product can be expressed by

$$gD = g_m d - \alpha(D - d),\qquad(5.143)$$

where, $g_m = \sigma_e(\lambda_m)N_2 - \sigma_a(\lambda_m)N_1$ is the gain coefficient at the mode's wavelength λ_m inside the pumped volume, assuming homogeneous populations densities within the disk, and $\alpha = \sigma_a(\lambda_m)N_1$ the absorption coefficient of this wavelength in the unpumped part of the disk.

As any kind of optical coating cannot change the total-internal reflection property of an optical medium with respect to its surrounding refractive index, the transverse modes will only be efficiently reflected at the HT side for $\theta_a > \theta_c$. Thus a simultaneous total-internal reflection on the cylindrical surface can only occur at angles of

$$\theta_c < \theta_a < 90° - \theta_c.\qquad(5.144)$$

For YAG, this results in $33.3° < \theta_a < 56.7°$. If we now assume that all the reflections are perfect, it follows directly from Eq. (5.143) and Eq. (5.141), that in a four-level-laser medium ($\alpha = 0$) the parasitic modes will immediately oscillate as $g_m d \geq 0$, and no laser action will occur through the cavity modes. For a quasi-three-level laser medium ($\alpha > 0$), we can rewrite the parasitic mode threshold $gD \geq 0$ using $N_1 + N_2 = N_{tot}$ as

$$\frac{N_2}{N_{tot}} \geq \frac{\sigma_a(\lambda_m)}{\sigma_e(\lambda_m) + \sigma_a(\lambda_m)} + \left(\frac{D}{d} - 1\right)\sigma_a(\lambda_m).\qquad(5.145)$$

As a result of the McCumber relation in Eq. (2.32) the absorption cross section decreases exponentially with respect to the emission cross section at the long wavelength edge of the emission spectrum. Therefore, Eq. (5.145) can be always fulfilled for a given N_2 by a long-enough wavelength of the parasitic mode.

In the case of a non-perfect reflection, the condition for the onset of parasitic lasing, Eq. (5.141), results in a minimum reflection coefficient

$$R_k(\theta_a) = R_{HR}^k(\theta_a) R_{HT}^k(\theta_a) R_C(\theta_a),\qquad(5.146)$$

which is given by

$$\ln R_k(\theta_a) \geq \frac{N_{tot}}{\sin\theta_a} D\left(\sigma_a(\lambda_m) - \frac{d}{D}\frac{N_2}{N_{tot}}\left(\sigma_e(\lambda_m) + \sigma_a(\lambda_m)\right)\right).\qquad(5.147)$$

In the case of a perfect reflection on the HR and HT sides and $R_C(\theta_a) < 1$, we obtain a maximum gain for $\theta_a = \theta_c$. Then, the minimum reflection of the cylindrical surface is obtained as

$$\ln R_C(\theta_c) \geq N_{tot} n D\left(\sigma_a(\lambda_m) - \frac{d}{D}\frac{N_2}{N_{tot}}\left(\sigma_e(\lambda_m) + \sigma_a(\lambda_m)\right)\right).\qquad(5.148)$$

It can be shown that the minimum reflection $R_C(\theta_a)$ in this case can be very small, resulting in the need for a strong control of this value in order to suppress the transverse modes. In the other important case when a non-perfect reflection $R_{HR}(\theta_a) < 1$

Fig. 5.55 Suppression of transverse and radial modes by a non-cylindrical disk surface

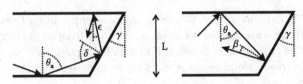

is assumed at the HR side, whilst all other sides are perfectly reflecting, we obtain a maximum gain for $\theta_a = 90° - \theta_c$. This modes shows the lowest number of reflections on the HR side. The minimum reflectivity then results in

$$\ln R_{HR}\left(90° - \theta_c\right) \geq 2nLN_{tot}\left(\sigma_a(\lambda_m) - \frac{d}{D}\frac{N_2}{N_{tot}}\left(\sigma_e(\lambda_m) + \sigma_a(\lambda_m)\right)\right). \quad (5.149)$$

In CW Yb^{3+}:YAG disk lasers this value can be as high as $R_{HR}(\theta_a) \sim 0.99$ without causing transverse modes to oscillate, whilst in Q-switched systems with a higher inversion density, the reflectivity of the HR mirror in the angular range discussed here needs to be low enough, e.g. $R_{HR}(\theta_a) < 0.8$ to suppress the transverse modes. A low HR reflectivity, however, results in a larger amount of fluorescence leaking through the HR mirror. This fluorescence gets absorbed by the heat sink and thus increases the cooling power necessary in order to maintain the disk temperature. As can also be seen in Eq. (5.149), a short crystal length L is beneficial to increase the oscillation threshold of the transverse modes. However, this has to be compensated for by a larger number of pump passes.

The radial modes do not show any total-internal reflection and oscillate through the disk center on the Fresnel reflection provided by the cylindrical disk surface, i.e. at $\theta_a = 90°$. They usually do not occur in cw lasers as a result of the low inversion density in cwW operation and the corresponding low gain. However, radial modes can occur in Q-switched disk lasers which show a much higher inversion density. The threshold for the radial modes can thus be expressed by

$$\ln R_C\left(90°\right) \geq N_{tot}D\left(\sigma_a(\lambda_m) - \frac{d}{D}\frac{N_2}{N_{tot}}\left(\sigma_e(\lambda_m) + \sigma_a(\lambda_m)\right)\right). \quad (5.150)$$

To suppress the transverse and radial modes, the outer disk surface can be shaped in a conical form shown in Fig. 5.55. Starting with an incident angle on to the bottom surface of θ_a the mode will be reflected from the outer surface and its incidence angle on the top surface will be

$$\epsilon = |\theta_a - 2\gamma|. \quad (5.151)$$

The mode is thus non-total-internal reflecting for

$$\frac{90° - \theta_c}{2} < \gamma < \theta_c, \quad (5.152)$$

which can only be fulfilled for $n < 2$, e.g. for YAG, where $28.4° < \gamma < 33.3°$. A mode reflecting totally on the upper surface needs a cone angle γ of

$$90° - 2\theta_c < \gamma < \theta_c \quad (5.153)$$

to be non-total-internal reflecting on the bottom surface, resulting in an incident angle of

$$\beta = |\, 90° - \gamma - \theta_a \,|. \tag{5.154}$$

For $n < 2$, this relation is also fulfilled when Eq. (5.152) is fulfilled. Thus, for YAG, a cone angle in the range $28.4° < \gamma < 33.3°$ has to be chosen for transverse and radial mode suppression.

References

1. F.K. Kneubühl, M.W. Sigrist, *Laser* (Teubner, Stuttgart, 1999)
2. M. Eichhorn, S.D. Jackson, Appl. Phys. B **90**, 35 (2008)
3. W. Koechner, *Solid-State Laser Engineering* (Springer, Berlin, 1999)
4. M. Monerie, Y. Durteste, P. Lamouler, Electron. Lett. **21**, 723 (1985)
5. S.A. Payne, L.L. Chase, L.K. Smith, W.L. Kway, W.F. Krupke, IEEE J. Quantum Electron. **28**, 2619 (1992)
6. A.A. Kaminskii, *Laser Crystals*. Springer Series in Optical Science, vol. 14 (Springer, Berlin, 1990)
7. J.B. Gruber, M.E. Hills, R.M. Macfarlane, C.A. Morrison, G.A. Turner, G.J. Quarles, G.J. Kintz, L. Esterowitz, Phys. Rev. B **40**, 9464 (1989)
8. C.A. Morrison, R.P. Leavitt, Spectroscopic properties of triply ionized lanthanides in transparent host crystals, in *Handbook on the Chemistry and Physics of Rare Earths*, ed. by K.A. Gschneider Jr. (North-Holland Publishing Co., Amsterdam, 1982)
9. B.M. Walsh, N.P. Barnes, B. Di Bartolo, J. Appl. Phys. **83**, 2772 (1998)
10. S. Hufner, *Optical Spectra of Transparent Rare Earth Compounds* (Academic Press, New York, 1978)
11. B. Henderson, G.F. Imbusch, *Optical Spectroscopy of Inorganic Solids* (Clarendon Press, Oxford, 1989)
12. A. Richter, Ph.D. thesis, University of Hamburg, Germany, 2008
13. M. Alshourbagy, Ph.D. thesis, University of Pisa, Italy, 2005
14. K.M. Dinndorf, Ph.D. thesis, Massachusetts Institute of Technology, USA, 1993
15. T. Förster, Ann. Phys. **437**, 55 (1948)
16. T. Förster, Z. Naturforsch. A **4a**, 321 (1949)
17. D.L. Dexter, J. Chem. Phys. **21**, 836 (1953)
18. C.Z. Hadad, S.O. Vasquez, Phys. Rev. B **60**, 8586 (1999)
19. P.M. Levy, Phys. Rev. **177**, 509 (1969)
20. F.R.G. de Silva, O.L. Malta, J. Alloys Compd. **250**, 427 (1997)
21. V.S. Mironov, J. Phys. Condens. Matter **8**, 10551 (1996)
22. M. Yokota, O. Tanimoto, J. Phys. Soc. Jpn. **22**, 779 (1967)
23. A.I. Burshtein, Sov. Phys. JETP **35**, 882 (1972)
24. T. Holstein, S.K. Lyo, R. Orbach, in *Topics in Applied Physics*, vol. 49, ed. by W.M. Yen, P.M. Selzer (Springer, New York, 1981). Chap. 2
25. R. Orbach, in *Optical Properties of Ions in Crystals*, ed. by H.M. Crosswhite, H.W. Moos (Interscience, New York, 1967), p. 445
26. S. Xia, P.A. Tanner, Phys. Rev. B **66**, 214305 (2002)
27. S.D. Jackson, Opt. Commun. **230**, 197 (2004)
28. A. Hoffstaedt, Festkörper-Laser-Institut Berlin, Germany, 1991
29. M. Rattunde, Ph.D. thesis, University of Freiburg i. Br, Germany, 2003
30. D. Gloge, Weakly guiding fibers. Appl. Opt. **10**, 2252 (1971)
31. B.M. Walsh, N.P. Barnes, Appl. Phys. B **78**, 325 (2004)

32. B.M. Walsh, Private communication
33. M. Eichhorn, A. Hirth, in *Conference on Lasers and Electro-Optics CLEO 2008*, San Jose, USA. Paper CTuII3
34. M. Eichhorn, Opt. Lett. **32**, 1056 (2007)
35. M. Eichhorn, in *Conference on Lasers and Electro-Optics CLEO 2007*, Baltimore, USA. Paper CTuN7
36. M. Eichhorn, S.D. Jackson, Opt. Lett. **32**, 2780 (2007)
37. M. Eichhorn, S.D. Jackson, in *Conference on Lasers and Electro-Optics CLEO 2008*, San Jose, USA. Paper CFD7
38. M. Eichhorn, in *OPTRO 2005 Symposium*, Ministère de la Recherche, Paris, France, 9–12 May 2005
39. M. Eichhorn, in *Journées Scientifiques de l'ONERA: Lasers et amplificateurs à fibre optique de puissance: fondements et applications*, ONERA, Châtillon, France, 27–28 June 2005
40. M. Eichhorn, Opt. Lett. **30**, 456 (2005)
41. M. Eichhorn, Opt. Lett. **30**, 3329 (2005)
42. M. Eichhorn, Virtual J. Ultrafast Sci. **5**(1) (2006)
43. G.P. Agrawal, *Nonlinear Fiber Optics* (Academic Press, San Diego, 2001)
44. D. Cotter, J. Opt. Commun. **4**, 10 (1983)
45. L.G. Hwa, J. Schroeder, X.S. Zhao, J. Opt. Soc. Am. B **6**, 833 (1989)
46. Y. Durteste, M. Monerie, P. Lamouler, Electron. Lett. **21**, 723 (1985)
47. K. Contag, Dissertation, University of Stuttgart, Germany, 2002
48. D. Marcuse, Bell Syst. Tech. J. **55**, 703 (1977)
49. S.P. Timoshenko, J.N. Goodier, *Theory of Elasticity*, 3rd ed. (McGraw-Hill, New York, 1970)

Index

A
Absorption, 2, 6
Absorption-cross section, 15, 18
Absorption-averaged beam radius, 119
Acceptance solid angle, 141
Acoustic phonons, 154, 156
Acousto-optic figure of merit, 82
Acousto-optic modulator, 71, 81, 94, 151
   external diffraction angle, 83
   internal deflection angle, 83
Active mode locking, 94
Active Q-switching, 75
Amplified spontaneous emission (ASE), 149, 150
Anti-Stokes process, 154
Arc length, 113
ASE threshold, 152
Azimuthal mode number, 143

B
Beam quality, 65
Beam quality factor, 66
   effective wavelength, 66
Bessel differential equations, 143
Boltzmann distribution, 24
Bose-Einstein distribution, 11
Bragg angle, 83
Bragg gratings, 148
Bragg scattering, 83
Brewster setup, 64
Brillouin amplification, 155
Brillouin line width, 155
Brillouin pump radiation, 154
Brillouin scattering, 154
Brillouin threshold, 155
Brillouin-scattering rate, 154
Build-up time, 79

Bulk pulse compression, 100

C
Casimir force, 13
Cavity dumping, 85
Cavity resonance, 70
Chemical potential wavelength, 31
Chirp, 93
Chirped-pulse amplification, 102
CO_2 laser, 105
Coherence, 46
Coherent photon number, 74
Complex beam radius, 57
Configurational coordinate model, 132
Conjugate variables, 66
Continuous wave (cw), 35
Coulomb interaction, 28
Coupling parameter, 127
Cr^{2+}:ZnSe laser, 64, 135
Cr^{3+}:LiCaAlF$_6$ laser, 135
Cr^{3+}:LiSrAlF$_6$ laser, 135
Cross relaxation, 127
Cross section
   spectroscopic, 31
Cross sections, 14
Cross-relaxation process, 121
Crystal field, 30
Crystal growth, 108
Crystal-field splitting, 131
Czochralski method, 108, 133

D
Damage threshold, 154
Damped harmonic oscillator, 42
Diffraction grating, 101
Diffraction loss, 67, 69
Diffraction orders, 82

M. Eichhorn, *Laser Physics*, Graduate Texts in Physics,
DOI 10.1007/978-3-319-05128-4,
© Springer International Publishing Switzerland 2014

Printed in the United States
By Bookmasters